MW00826121

Aristotle

Aristotle

FROM ANTIQUITY TO THE MODERN ERA

Preface by
MARTIN J. GROSS

Essays by
BENJAMIN MORISON and BARBARA SCALVINI

MARTIN J. GROSS FAMILY FOUNDATION, Livingston, New Jersey
In association with
D GILES LIMITED

Dedicated to Jordan and Benjamin Altman
and to my wife, Ahuva, their Safta

Published on the occasion of the exhibition *Aristotle: From Antiquity to the Modern Era* on display at the New-York Historical Society from March 26, 2021 until July 4, 2021

First published in 2021 by GILES
An imprint of D Giles Limited
66 High Street,
Lewes, BN7 1XG, UK
gilesltd.com

ISBN Hardcover: 978-1-911282-75-4

Library of Congress Control Number: 2020910094

For the Martin J. Gross Family Foundation:
Project Curator: Barbara Scalvini
Photographs: Ardon Bar-Hama

For D Giles Limited:
Copy-edited and proof-read by Jodi Simpson
Designed by Helen Swansbourne

Produced by GILES, an imprint of D Giles Limited
Printed and bound in Slovenia

Front cover: Columbanus Parent and Antonius Chassé (17th century), *Octo libri Physicorum* ..., France, 1664–65 (detail of cat. 27)

Back cover: Aristotle, *Opera*, end of *Prior Analytics*, Venice: Aldus Manutius, 1495–98 (detail of cat. 1)

Frontispiece: *Bust of Aristotle*, c. 320 BCE, marble, Roman copy after Greek original, Kunsthistorisches Museum, Vienna. Alamy Photo

Page 40: Aristotle and Robert Grosseteste (1175–1253), *Ethica ad Nicomacum*, Northern Italy, c. 1425 (detail of cat. 31)

CONTENTS

CATALOGUE

PREFACE

ARISTOTLE OF STAGIRA (384–322 BCE) is one of the most important philosophers of the ancient world, if not of all time. More than anyone, he set the stage for the structure and terms of Western philosophical thought. Of his surviving dialogues, presumably written under the influence of his teacher Plato, only fragments survive. What we consider to be the body of his teaching—the Aristotelian corpus—is derived from what we believe to be surviving notes of his lectures given at the Lyceum in Athens. The story of how his teachings have come down to us over the millennia is complex and fascinating. Helping to tell that story has been a rewarding intellectual journey.

What was known as Aristotelianism evolved over the ages as Aristotle's works gradually became available. His works on logic, for example, were known in the West centuries before his works on physics. It was not until the end of the twelfth century that many of his works reached Western scholars trained in Latin through translations and commentaries by people like Robert Grosseteste, first chancellor of Oxford University. Only by 1495–98, close to two millennia after his death, did the known Aristotelian corpus become available in Greek, thanks to Aldus Manutius's Aldine Press in Venice.

In the centuries before the advent of printing scribes copied scarce and fragile manuscripts; translators worked off these manuscripts, often producing textual variations; and commentators approached their often limited source material anchored in their respective religious and philosophical traditions. These scribes, translators, and commentators lived in different places at different times and conversed in various languages, primarily Greek, Latin, Arabic, Hebrew, and Syriac, as well as many others. The hunt for precious manuscripts was never ending.

Two prominent scholars set the stage for the preservation and study of the Aristotelian corpus. The first, known as "The Commentator," was Alexander of Aphrodisias, who flourished around 200 CE. Without his commentaries, little of the Aristotelian corpus might have survived. The second was Ibn Rushd, known as Averroes, the twelfth-century Muslim commentator, who exercised a significant influence on Latin scholastic thinkers from the thirteenth century onwards. While Averroes was reconciling Islamic teaching with Aristotelian doctrine, his great Jewish contemporary and fellow Cordovan Moses Maimonides relied heavily on Aristotelian doctrine to harmonize Jewish teachings with Greek philosophical thought. A century later, Thomas Aquinas did something similar with Christianity.

The Aristotelian publications featured here traverse three main periods: the medieval, the Renaissance, and the early modern. The medieval period encompasses the reception of Aristotle's teachings by Latin-speaking writers in the West from 500 CE to around 1450; the Renaissance runs from around 1450 through the end of the sixteenth century; and the early modern period covers the sixteenth–eighteenth centuries. Each book contributes to our understanding of the reception, interpretation, and transmission of the Aristotelian corpus over the centuries.

The exhibition that brought these works together had its origin in a collection of books and manuscripts that, like most collections, resulted from a combination of happenstance and personal interest. While I have been collecting philosophy and history of science books for several decades, this particular collection of Aristotle-related books and manuscripts began during a business trip to London a few years back. I had some time between meetings, and it turned out that Bernard Quaritch Ltd., the antiquarian bookstore, was around the corner from where I was staying. Quaritch has been dealing in rare books and manuscripts for almost two hundred years. Having done business with them previously, I wandered into the shop to browse, where I met Barbara Scalvini, one of the staff. On the table was a book she had recently purchased at auction. While I took an immediate interest in it, Barbara told me that it had been reserved by someone else. Fortunately for me, the other party could not find the funds for it, so I was able to acquire the first book of my collection of editions of Aristotle, commentaries on Aristotle's works, lecture notes for courses on Aristotle, and books and manuscripts by or on Aristotle with significant marginal annotations in them. Since that initial purchase, my collection has grown to over forty carefully selected titles under Barbara's discerning eyes. For an object to make its way into the collection, it needs to be relevant and uncommon. And it must help tell part of the story of the transmission of the thought of Aristotle from antiquity to the modern era.

That first book in particular was fascinating for several reasons. Published in 1534 and bound in contemporary vellum, it was the second edition of Johannes Philoponus's commentary on Aristotle's *Posterior Analytics*, a foundational work on scientific method. It was published by the Aldine Press, which played a major role in the dissemination of classical texts throughout Europe. Interestingly, this copy had two unique aspects to it: the first was the fact that it contained over one hundred annotations in the margins of the text; the second was that its prior owner, Francesco Buonamici, was one of Galileo's professors at Pisa. A fascinating question presented itself: Could there be contained in the annotations a demonstrable link between Aristotle's account of scientific method and Galileo's views on the universe?

This one item suggested many of the themes of the exhibition and this accompanying catalogue: the substance of some of the most important doctrines of Aristotle, the Aldine Press and its role in spreading knowledge of the classics, and the importance and role of annotations. The issues are complex: How did the thought of someone who lived more than two millennia ago get handed down across time, space, and language? How were Aristotle's works translated? How were they commented on? How did they survive?

To attempt to tell such a nuanced story is a challenge. To do so in a way that is accessible and of interest to a general audience is doubly challenging. To help set a context, Benjamin Morison, Professor of Philosophy at Princeton University, wrote a magnificent essay on the philosophy of Aristotle. Barbara Scalvini used her training in early modern European thought to write the object labels for the exhibition, and she also contributed a wonderful essay on the different ways in which the Aristotelian corpus was transmitted.

For those in whom this instills an even greater thirst for knowledge about this subject, suggestions for further reading are included in the back of the catalogue. Needless to say, for some this will be too large a chunk to digest, for others just the beginning of a wonderful journey. Each person is invited to use this catalogue as an introduction to a world of serious study central to the intellectual history of the West. I am very grateful to Ben and Barbara for the time and effort they have spent to make this exhibition and its accompanying catalogue possible. I would also like to thank Ardon Bar-Hama for his meticulous work photographing objects for the exhibition. Finally, I want to thank Louise Mirrer, President and CEO of the New-York Historical Society, along with its Board of Trustees, for allowing this exhibition to take place at the Society. Last but not least, this exhibition could never have happened without the invaluable guidance and great attention paid to it by Michael Ryan, the Vice President and Sue Ann Weinberg Director of the Patricia D. Klingenstein Library at the New-York Historical Society.

Finally, a personal note. A number of people have asked me why I spent the time collecting these books and manuscripts, and putting together this exhibition and catalogue. Why should anyone be interested in what Aristotle, a man who lived over two thousand years ago, had to say? The short answer is that an appreciation of ancient philosophy is essential to having a basic understanding of Western civilization. Civilizations are defined by many things, especially the questions they pose to themselves. The Bible and ancient Greek philosophy played a key role in setting the stage for the West: religious inspiration and philosophical inquiry; Jerusalem and Athens; God and man—to drastically oversimplify it. Many of the foundational questions the West continues to grapple with were first posed by the Greek philosophers. What is the nature of being? What is it to know something and what is it that is known—physical things, mathematical objects, abstract ideas? What is the nature of the physical world? What are the forces that act within it? How should we think about concepts like matter, form, place, time, change, and many others? How, and when, did it come to be? Or was it always there? Will it end one day? What is the nature of mathematics? What are the rules of logical inference? How should people behave and what ultimate goals should we seek? How should society organize itself? Aristotle himself wrote a Constitution for Athens. Perhaps most important, is it possible for limited and finite beings like ourselves to attain a glimpse of or participate in something more transcendent and timeless? And many more.

As you will see from the essays herein, Aristotle was a truly pivotal figure. He has been called the last of the ancients and the first of the moderns. To employ Aristotle as the pathway to investigating these foundational questions could not be more appropriate. If this exhibition and catalogue help the viewer achieve a greater appreciation for some of the foundational questions on which Western civilization is based, then it will more than achieve its purpose.

MARTIN J. GROSS
Livingston, NJ

WORKS OF ARISTOTLE

In *The Complete Works of Aristotle: The Revised Oxford Translation*, edited by Jonathan Barnes.

- Categories
- De Interpretatione
- Prior Analytics
- Posterior Analytics
- Topics
- Sophistical Refutations
- Physics
- On the Heavens
- On Generation and Corruption
- Meteorology
- On the Soul
- Sense and Sensibilia
- On Memory
- On Sleep
- On Dreams
- On Divination in Sleep
- On Length and Shortness of Life
- On Youth, Old Age, Life and Death, and Respiration
- History of Animals
- Parts of Animals
- Movement of Animals
- Progression of Animals
- Generation of Animals
- Metaphysics
- Nicomachean Ethics
- Magna Moralia
- Eudemian Ethics
- Politics
- Rhetoric
- Poetics
- Constitution of Athens

γ οὐδενὶ· ὥστε τὸ α τῷ γ οὐδενί· οὐκ ἄρα ἀναπνεῖ ὁ τοῖχος.

Ἐοίκασι δὲ αἱ τοιαῦται τῶν αἰτιῶν, τοῖς καθ' ὑπερβολὴν εἰρημένοις· τοῦτο δ' ἐστὶ τὸ πλέον ἀποστήσαντα τὸ μέσον εἰπεῖν· οἷον τὸ τοῦ Ἀναχάρσιδος, ὅτι ἐν Σκύθαις οὐκ εἰσὶν αὐλητρίδες· οὐδὲ γὰρ ἄμπελοι. κατὰ μὲν δὴ τὴν αὐτὴν ἐπιστήμην, καὶ κατὰ τὴν τῶν μέσων θέσιν, αὗται διαφοραί εἰσι τοῦ ὅτι πρὸς τὸν διότι συλλογισμόν. ἄλλον δὲ τρόπον διαφέρει τὸ διότι τοῦ ὅτι, τὸ δι' ἄλλης ἐπιστήμης ἑκάτερον θεωρεῖν. τοιαῦτα δ' ἐστὶν, ὅσα οὕτως ἔχει πρὸς ἄλληλα ὥστ' εἶναι θάτερον, ὑπὸ θάτερον· οἷον τὰ ὀπτικὰ, πρὸς γεωμετρίαν· καὶ τὰ μηχανικὰ πρὸς στερεομετρίαν· καὶ τὰ ἁρμονικὰ πρὸς ἀριθμητικήν· καὶ τὰ φαινόμενα πρὸς ἀστρολογικήν. σχεδὸν δὲ συνώνυμοί εἰσι τούτων τῶν ἐπιστημῶν ἔνιαι· οἷον ἀστρολογία ἥ τε μαθηματικὴ καὶ ἡ ναυτική· καὶ ἁρμονικὴ ἥ τε μαθηματικὴ, καὶ ἡ κατὰ τὴν ἀκοήν· ἐνταῦθα γὰρ τὸ μὲν ὅτι, τῶν αἰσθητικῶν εἰδέναι· τὸ δὲ διότι, τῶν μαθηματικῶν· οὗτοι γὰρ ἔχουσι τῶν αἰτίων τὰς ἀποδείξεις· καὶ πολλάκις οὐκ ἴσασι τὸ ὅτι, καθάπερ οἱ τὸ καθόλου θεωροῦντες, πολλάκις ἔνια τῶν καθ' ἕκαστον οὐκ ἴσασι δι' ἀνεπισκεψίαν. ἔστι δὲ ταῦτα ὅσα ἕτερόν τι ὄντα τὴν οὐσίαν κέχρηται τοῖς εἴδεσι. τὰ γὰρ μαθήματα, περὶ εἴδη ἐστίν, οὐ γὰρ καθ' ὑποκειμένου τινός· εἰ γὰρ καὶ καθ' ὑποκειμένου τινὸς τὰ γεωμετρικά ἐστιν, ἀλλ' οὐχ ᾗ γεωμετρικά, καθ' ὑποκειμένου. ἔχει δὲ καὶ πρὸς τὴν ὀπτικὴν, ὡς αὕτη πρὸς τὴν γεωμετρίαν, ἄλλη πρὸς ταύτην· οἷον τὸ περὶ τῆς ἴριδος· τὸ μὲν γὰρ ὅτι, φυσικοῦ εἰδέναι· τὸ δὲ διότι, ὀπτικοῦ· ἢ ἁπλῶς ἢ κατὰ τὸ μάθημα. Πολλαὶ δὲ καὶ τῶν μὴ ὑπ' ἀλλήλας ἐπιστημῶν, ἔχουσιν οὕτως, οἷον ἰατρικὴ πρὸς γεωμετρίαν· ὅτι μὲν γὰρ τὰ ἕλκη τὰ περιφερῆ βραδύτερον ὑγιάζεται, τοῦ ἰατροῦ εἰδέναι· διότι δὲ, τοῦ γεωμέτρου. Τῶν δὲ σχημάτων, ἐπιστημονικὸν μάλιστα, τὸ πρῶτόν ἐστιν· αἵ τε μαθηματικαὶ τῶν ἐπιστημῶν διὰ

Aristotle

An Introduction

BENJAMIN MORISON

It's easy to make the case that Aristotle was the greatest philosopher who ever lived. I offer three pieces of evidence. First, the staggering range of his works. He worked in every conceivable field of philosophy: logic—check; epistemology—check; metaphysics, ethics, political theory, philosophy of language, rhetoric, natural science, aesthetics, psychology, economics, astronomy—check, check, check, check ... But Aristotle didn't just work in those fields; he invented several of them. There had never been a treatise on logic before the *Prior Analytics*, as Aristotle himself, with characteristic bluntness, was keen to point out to the reader (modesty does not figure in the list of virtues in his *Nicomachean Ethics*). His teacher Plato—and before Plato, Socrates—had argued with beautiful precision and subtlety, but Aristotle attempted to codify and understand the workings of the distinctive patterns of discursive reasoning. And having succeeded (this is a man who didn't often fail, though we will hear about some of his failures later), he promptly proceeds to offer the reader advice as to how to do it well. In the field of natural science, too, Aristotle was quite simply revolutionary. Where Plato was pessimistic about the possibility of gaining knowledge of the physical world, Aristotle saw that the imperfections and apparent chaos of the world around us could be tamed and brought under theoretical control. In many of the preexisting fields Aristotle worked in, Plato had made a good start,

Aristotle
Posterior Analytics
Venice: Aldus Manutius, 1495–98
(detail of cat. 1)

philosophically speaking, especially in metaphysics, ethics, epistemology, and political philosophy. Aristotle took up the challenges set by Plato's suggestive, elusive, and fertile dialogues and in response came up with detailed, intricately argued theories, laid out as philosophical treatises, not dialogues. As often as not, he found fault with his teacher, as in the case of natural science, famously implying that while Plato might have been his friend, truth is more of a friend. (He must have been a nightmare to have as a student—a fact missed by those later philosophers who could not bring themselves to believe that Plato and Aristotle disagreed on anything.) In short, then, our first reason to be in awe of his philosophical legacy is the sheer range of his philosophizing.

Second, Aristotle is still a living presence in the philosophical community. In every undergraduate philosophy curriculum his works will be studied; in most university philosophy departments there will be one resident Aristotelian, or maybe more (I am one of two card-carrying Aristotelians in mine). In every journal of philosophy his texts are still being interpreted and argued over by scholars, some of whom learned Greek just in order to read him. Philosophers forging their own theories and systems of thought still treat Aristotle's theories as either welcome sources of inspiration or dangerous alternatives in need of refutation.

Third, Aristotle's works have been more or less continuously studied since he wrote them, some two and a half thousand years ago. Admittedly, there was a rocky start. Immediately after his death, things were in full swing: he had pupils who took up his mantle, continuing the extraordinary research into the natural world that he himself had inaugurated (more on this below). But over the next couple of hundred years, interest in his works dwindled. Later historians and philosophers hypothesized that the libraries containing his works had been pillaged and the last remaining copies of his works scattered, stashed away in caves by true believers for safekeeping. It's a story, of course, but Aristotle himself loved a good story: the lonelier he got in old age, the more he is said to have turned to myths. You can see why such a story needed to be invented. How else could it be that the man who later would be known simply as "The Philosopher" failed to exercise influence in the centuries immediately after his death?

Fortunately, although interest in Aristotle dwindled, it never disappeared. Around the turn of the millennium, a definitive edition of his works was published, commentaries on his works started to appear, and once again people were self-identifying as Aristotelians, or Peripatetics: just as Plato had his Academy, giving rise to the word *Academics* for those who followed him, Aristotle's school in Athens, the Lyceum, had been associated with the *peripatos*, a covered colonnaded walkway, up and down which Aristotle apparently used to walk while lecturing. From this point on, his works take their place in the philosophical curriculum. Not all his works were studied all the time—in the Middle Ages in the West, there was a rekindling of interest in his physics and metaphysics, and a corresponding increase in the availability of those texts. But as far as we can tell, his *Categories* and *De Interpretatione* have been read continuously from the first century CE.

Of course, the real proof of Aristotle's greatness lies in the philosophical power of his thought. But before we examine this in more detail, it is worth taking a step back in time with a short history of the philosophy that predated Aristotle. This philosophical tradition conventionally begins with Thales, a philosopher working at the start of the sixth century BCE in Miletus, an Ionian colony in Asia Minor. None of Thales's works survive (a situation all too common with the philosophers who predate Socrates, and who are known collectively as the Presocratics), but stories about Thales abound. He was a keen astronomer, and famously predicted an eclipse; he was mocked by a maid for falling down a well while walking with his eyes gazing at the stars above him. This is the beginning of a long-standing tradition of depicting philosophers as having their heads in the clouds, or being out of touch with the world around them—indeed, the great comic playwright Aristophanes will later write a play called *The Clouds*, targeting Socrates. In Thales's case, this characterization seems to be particularly unfair: the source of the story is probably the fact that the ancients believed the stars could be seen in the daytime from the bottom of wells; I expect Thales was trying this out, and got mocked for it. Moreover, Thales was an especially canny operator: one year, predicting a good olive harvest (no doubt through some nifty astronomical observations), he bought up all the local olive presses, and then made a killing on olive oil when his prediction proved correct.

For our purposes, though, Thales is important because of a doctrine attributed to him by Aristotle—though he does not seem to have had access to Thales's works, and is quite tentative in his attributions. That doctrine is that everything is made out of water, and comes from water. This remarkable claim constitutes an attempt to understand the natural world around us. Thales thought the features of our world could be explained by the fact that it is made of water: it is fluid and in motion all the time; things form and reform, condense and rarefy, move from soft to brittle, or evaporate entirely away, as water, ice, and steam do. Other thinkers, such as Anaximander and Anaximenes, attempted similar analyses, positing other fundamental principles: maybe the cosmos is made out of air, not water; or maybe it is made of fire. What they all have in common is that they no longer rely solely on myth to understand the world; they want to explain and understand the world in a different mode, using the tools of reason and perception.

But reason and perception are not always allies. The attempt at natural science and explaining the cosmos came under suspicion from philosophers such as Parmenides and Zeno, who worked in the first part of the fifth century BCE. Both hailed from Elea, in Italy, the opposite side of the Greek-speaking world from Thales. They belonged to the school known as the Eleatics, who argued that the very phenomenon the Milesians were trying to explain—the fluid and changing cosmos around us—is not real, because change is impossible. Zeno's famous paradoxes of motion are good examples of the kind of reasoning that can lead to such a belief. One of these paradoxes, known as "Achilles and the Tortoise," runs (or doesn't, as the case might be) like this: The famously swift Greek hero Achilles and a tortoise are running a race. Achilles generously gives the tortoise a head start. Let us call the place where the tortoise starts T1. The race starts, and Achilles quickly catches up to T1, where the tortoise started. But in the time it took Achilles to reach T1, the tortoise has moved on to a new position, T2. Achilles arrives at T2 only to find the tortoise has moved on a little further, to T3. Achilles runs on to T3, but the tortoise is now at T4, and so on. Achilles can never catch the tortoise. Of course, we know from experience that Achilles will eventually catch the tortoise. But reason (or this piece of reasoning, anyway) tells us he won't. Parmenides and Zeno exhort us to follow reason, not our faulty and deceptive perception. Don't be deceived into thinking there is motion or change in the world, they tell us; it is not really happening. It is like those popular optical illusions of motion that you see on the Internet all the time: the circles look like they are rotating (or whatever), but they aren't. Perception tells you they are; reason tells you they are not.

Out of this background emerges the towering figure of Socrates, towards the end of the fifth century BCE. He wrote nothing, and we only know about him because his pupils Xenophon and Plato dutifully recorded what we take to be faithful renditions (in spirit if not in letter) of his unique argumentative style. Socrates marks a new departure for philosophy, shifting the emphasis from natural philosophy (or its impossibility) to ethical questions, and in particular, the million-dollar question: How should I lead my life? Socrates seems to have been particularly struck by the fact that lots of people went around the place offering sage advice about such important questions, and yet these very same people all crumbled under the pressure of his sharp and agile questioning, and ended up contradicting themselves or espousing ridiculous and obviously incoherent positions. But how could someone really know what they were talking about if they contradicted themselves and didn't even have a coherent worldview? How could anyone like that be truly knowledgeable? Instead, Socrates proclaims, we must work towards clear and cogent definitions of the things that we value in our lives, such as piety, courage, friendship, and virtue, so that we can conduct rational investigations using them as our starting points. Socrates recognized that he didn't yet know of any such rigorous definitions, and therefore refused to claim that he was knowledgeable or wise—although he did point out that at least he knew he didn't possess the requisite kind of knowledge, unlike those whose authority he dedicated himself to puncturing.

Socrates's direct assault on the moral and epistemological authority of those around him earned him no friends, and he was put to death in

Raphael (1483–1520), *The School of Athens* (1509–11), detail of Plato (left) and Aristotle (right), fresco, Apostolic Palace, Vatican City. Alamy Photo

399 BCE on charges of impiety toward the traditional gods and corruption of the youth of Athens. But Socrates's legacy of asking the big questions about morality, and his insistence on the search for definitions of key notions on the basis of which one can conduct further investigation into those notions, lived on in his pupil Plato. Like Socrates, Plato did not engage in the project of natural science, comparing the material world to the world of dreams: fluid, unpredictable, impossible to pin down. But he conducted unparalleled investigations into the nature of justice, knowledge, and reality, concluding in the *Republic* that the world of perception is not the real world, but only a knock-off version of the ideal world of the Forms. Think of the famous Platonic Forms as being the things that are described by definitions. When we define a doctor as someone who knows how to cure people (to take an especially vivid example from book I of the *Republic*), we define an ideal that very possibly no doctor in the world around us actually embodies: all human doctors make mistakes or fall short in their knowledge. The true doctor, the real doctor, is the one who knows how to cure people, and such a doctor might not be possible in the material world. The attention that Plato lavishes on such idealized definitions prompted the Renaissance artist Raphael in his famous fresco *The School of Athens* (see opposite) to depict Plato as pointing up to the heavens, away from the world around us (again, that old trope of the philosopher with his head in the clouds).

This is where Aristotle comes in. We will take a look at three areas of Aristotle's philosophical system: logic, natural philosophy or natural science, and ethics. Each has been incredibly influential. Each has had its heyday at different times (in the twentieth and twenty-first centuries, Aristotle's work on ethics is riding particularly high). Each is, in its own way, startling, and each turns upside down that which came before it.

Logic

We have to start with logic. When the first edition of Aristotle's works was created in the first century BCE, the editor put Aristotle's logical works at the beginning of the set. Logic was thought of as the *organon*: the tool for doing philosophy and science. It was the first thing a student should master in their philosophical education.

Philosophy is all about reasoning. Reasoning is about discerning the connections between things. Facts don't exist in isolation from each other; they are, often, connected to one another. Here is a classic example. Take the following three statements:

1. All figures formed by joining the ends of the diameter of a circle to a random point on the circumference of that circle are triangles;
2. All triangles have angles that add up to 180°;
3. All figures formed by joining the ends of the diameter of a circle to a random point on the circumference of that circle have angles that add up to 180°.

The relationship between these three statements is such that if (1) and (2) are true, then (3) must also be true; it is impossible for (1) and (2) to be true without (3) being true. The statements conform to a pattern that guarantees this relationship between them. Statement (1) exhibits the pattern "All As are B"; (2) exhibits the pattern "All Bs are C"; and (3) exhibits the pattern "All As are C." Aristotle points out that any two statements that exhibit the pattern "All As are B" and "All Bs are C" will, if they are true, jointly guarantee the truth of the third statement, "All As are C," whatever A, B, and C might be. A, B, and C could belong to any field of enquiry; this pattern of reasoning holds everywhere.

Here is another instance of that same pattern, this time with three statements having nothing to do with geometry, as in the previous example:

1. All donuts are delicious;
2. All delicious things are bad for you;
3. All donuts are bad for you.

If the first two statements are true (maybe they are, maybe they aren't, but *if* they are true), then the third one must accordingly be true.

Aristotle describes this pattern of thought as a *syllogism*. He sets it out as a piece of reasoning:

All donuts (A) are delicious (B);
All delicious things (B) are bad for you (C);
Therefore, all donuts (A) are bad for you (C).

What marks this off as a piece of reasoning (as opposed to just three unrelated statements in a row) is the word *therefore* heralding the third statement. That word indicates that the last statement is being announced as being the conclusion of the piece of reasoning, with the statements preceding the word *therefore* being the assumptions, or premises, of the reasoning. Since this piece of reasoning is a syllogism, i.e. if the first two statements are true, the third must be, it is a peculiarly powerful intellectual tool: if you know that all As are B, and that all Bs are C, you would be in a position to know, because of that, that all As are C. You could move to this new thought, that all As are C, in perfect confidence that it must be true, given the other two things that you know. So imagine that you already know that all donuts are delicious (perhaps you have learned this through experience and extensive personal research). You then learn that all delicious things are bad for you (you make the rookie mistake of reading an online post about health). Now, you can put those two thoughts together and deduce that all donuts are bad for you. The two pieces of knowledge that you had come together to create a third, albeit unwelcome, piece of knowledge. (Whether you act on that knowledge is another matter. Aristotle will have much to say about this very human problem in his *Nicomachean Ethics*.)

Aristotle recognized the power of such structured reasoning, and asked himself the obvious question: What other forms or patterns of reasoning are like that? He came up with several other syllogisms. Here is another: All As are B; no B is C; therefore, no A is C. And here is another: All As are B; all As are C; therefore, some Bs are C. There are many more, and there are many patterns that may look as though they might be syllogisms but are not. What about this one, for instance: All As are B; all Bs are C; therefore, all Cs are A? Answer: No, it is not a syllogism. Aristotle, with his typically rigorous mind, sought to discover exhaustively not only which patterns make syllogisms and which do not, but also why each pattern succeeds or fails. The result is one of the most powerful presentations of systematic thought in the history of philosophy, enshrined in *Prior Analytics*, book I, chapters 4–7.

I am not the only one who admires this text. Aristotle himself was enamored of his creation and lauded his own achievement, remarking at the end of his *Sophistical Refutations* that it took him a long time to come up with the theory of the syllogism, and that we should pardon any omissions (he doesn't say *errors*), and thank him for his discoveries. The medieval monks who dutifully studied and recopied Aristotle's works were also smitten, and came up with a whole series of names for the syllogistic patterns to help them remember which patterns work and which ones don't. The most celebrated syllogism of all—All As are B; all Bs are C; therefore, all As are C—they lovingly named Barbara.

Logic is a tool for reasoning, but why should it be given pride of place, as the ancient editors of Aristotle's work insisted? The answer is that Aristotle was not interested in logic only because of its potential for extending our knowledge, as in the donut case above; he was also struck by the fact that you can systematize a body of knowledge using logic. These patterns of reasoning impose structure on groups of isolated facts: where before there were three thoughts sitting alongside each other in our mind, now we can discern that two of them can be used to deduce the third. The third is not, so to speak, equal to the other two; it is derived from them. For Aristotle, this was highly suggestive. He thought the world itself was also highly structured, with a set of basic units of knowledge serving to explain other things. Take the following facts about donuts: they are glazed fried batter; they are delicious. Aristotle would see here the potential to arrange these two facts into a kind of order: the fact that donuts are glazed fried batter explains why they are delicious. The deliciousness comes from the fact that they are glazed fried batter. Logic offers a wonderful way to regiment and systematize such explanatory relations. If we are to use our newfound logical skill to help us here, though, we need to incorporate some other facts to help bridge the gap between glazed fried batter and deliciousness. The following syllogisms show one way in which we might do this:

1. All donuts are glazed fried batter;
2. All glazed fried batter is fatty and sugary;
3. Therefore, all donuts are fatty and sugary.

3. All donuts are fatty and sugary;
4. All fatty and sugary things are delicious;
5. Therefore, all donuts are delicious.

Here, we have two syllogisms chained together; the conclusion of the first reappears as the opening premise of the second. Notice that we have, all in all, three statements about donuts: (1), (3), and (5). But statements (3) and (5) are both derived statements: (3) from (1) and (2), and (5) from (3) and (4). In fact, we used just one basic fact about donuts (along with other facts about batter, sugar, fat, and deliciousness) to derive the other two facts about donuts. That key fact is fact (1): donuts are glazed fried batter. This fact about donuts is a privileged one: it can be used to explain the others, using logic as the vehicle for our explanation. Aristotle's idea is that this privileged fact about an object—the one we will use to explain the other facts—should be the object's definition, stating its nature. The fact that a donut is glazed fried batter is a good candidate for being its definition: it tells us what donuts are, and serves as the basis for explaining other facts about donuts—that they are delicious, for instance, or bad for you, or shiny, fillable, etc.

In envisaging that we can explain everything about something by discovering its definition and then deploying that definition (in conjunction with other facts) as the basis for our explanations, Aristotle is channeling his teacher Plato (and Plato's teacher Socrates), with the distinctive Aristotelian logical twist that these explanations should be cast as syllogisms. But both Aristotle and Plato were influenced by the extraordinary achievements of the Greek geometers, of which Euclid's *Elements* is a wonderful representative (even though it dates from a little later than Aristotle). Book I of that work starts, in true Aristotelian fashion, with definitions (twenty-three of them) of the key items that will feature in the science—lines, circles, triangles, squares, etc.—and then sprinkles in a few more first principles (just ten of them), to produce a list of thirty-three starting points (premises) for the logical demonstrations that follow. In the *Posterior Analytics*, Aristotle produces his vision of how a science should be laid out with just such a stratification of facts in mind: the basic ones (including the definitions), and then the ones that follow from the basics. We call this second group of facts *theorems*. The Pythagorean theorem in geometry is called a theorem precisely because it is a derived fact, not one of the basic truths (in fact, it is the forty-seventh proposition to be proved in

Euclid's *Elements*, book I). So the geometers were already doing something of the sort that Aristotle recommends in the *Posterior Analytics*, probably without musing much about what they were doing; Aristotle codifies, analyses, streamlines, and packages up neatly the methodology that the geometers were using all along—using his new-found theory of logic as his guide. In a nutshell, logic is the tool we use for organizing and systematizing bodies of knowledge. There is no understanding without logic.

Natural Philosophy

If Aristotle's response to the geometers of his day was to codify their practice and philosophize about what they were up to, his next suggestion is positively world-shattering: What if all sciences—not just geometry—could be laid out this way, with definitions and general principles as starting points for the derivations of all the remaining truths that there are in each science? Plato thought that the only domains ripe for being known were highly theoretical, idealized domains of knowledge, such as mathematics and geometry, and the pure world of the Platonic Forms. But, thought Aristotle, what if the natural world around us was also susceptible to such treatment? What if knowledge of the natural world was just like geometry? What if Plato was wrong to think that the physical world cannot be known and understood by humans? Aristotle's conviction that this was so gave rise to an entire new field: natural philosophy or natural science, in the form of science as we think of it now, complete with the collection of essential data (empirical research), and the subsequent attempt to order that knowledge so as to find the definitions and basic principles from which theorems are derived.

Aristotle's most general and abstract foray into the domain of natural science is his *Physics*. In this work, Aristotle goes to the heart of all natural science and offers definitions of the key notions that it deals with: change, matter, form, nature, cause, chance, determinacy and indeterminacy, place, void, and time. When he offers these definitions, he also offers justifications for them: unlike Euclid, who just states with great authority what a triangle is, Aristotle will offer you an impressive battery of reasons for thinking his definition of place (for instance) is the right one. Because of that, the work

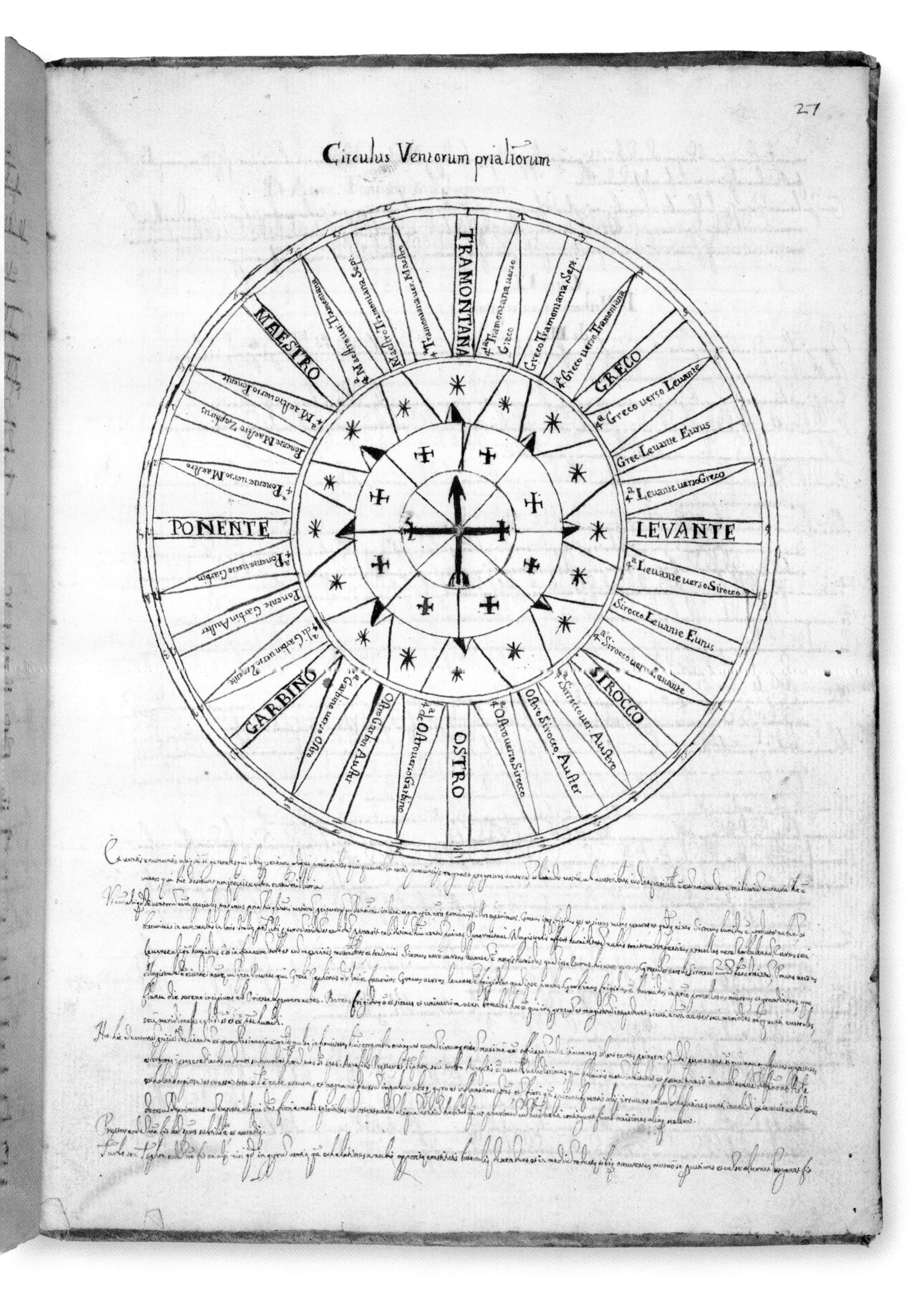

Giovanni Carlo Maffei (17th century)
In quatuor libri de coelo ad subtilium principis mentem
Northern Italy, 1677 (cat. 29); wind rose followed by microscript

reads to us more like a work of philosophy than of science. But the divide between philosophy and science is not a clear one for Aristotle. In fact, of all the pages that he wrote, about half of them fall under what we would call science, yet Aristotle obviously thinks of himself as doing philosophy when he is writing those works.

Aristotle pursues his guiding project ruthlessly in the scientific works. Take, for instance, human beings. Recall that Aristotle's claim is this: that we can find a fact about humans that will be able to explain everything in the field of human biology (in conjunction with facts about things other than humans). Now, we are distinctively rational creatures, employing reason in all areas of our lives, from planning the immediate minutiae of our everyday lives to deliberating about long-term matters such as money, family, and jobs; from writing history and poetry and philosophy to doing the most arcane and specialized scientific research. Humans are rational animals. Could we take this one fact about humans as a basic truth, and then derive all other facts about humans and what we are like from it?

There are two challenges this idea must overcome in order to be successful. First, will this definition of a human sit well alongside definitions of other animals? Make no mistake, Aristotle is aiming high here: he wants to come up with an account of all the different kinds of life, and that means finding a systematic way to describe plant life, animal life, and human life. He makes a simple suggestion: plants lead a life of nutrition, animals lead a life of perception, and humans lead a life of reason. Aristotle defines the bundle of capacities an organism has for leading its particular life as its soul: in *On the Soul*, he muses that the three different kinds of life are so different that the corresponding three types of soul are also radically different and not really unified as one kind of thing. Biology, then, must always come in three branches: human biology, zoology, and botany.

But the second challenge is more substantive: is it really possible to explain all of human anatomy, for instance, from the single fact that humans are rational creatures? Aristotle thinks it is. In fact, he offers this as a program for biological and zoological research. In *Parts of Animals*, he explicitly says that it is because of the nature of humans that they have the parts they have. An Aristotelian biological explanation of this claim might go something like this: Why do humans have large brains? Because all animals (including humans) must have a heart, since the heart (according to Aristotle) is the organ of perception. But hearts are hot, so the body needs a cooling mechanism. That cooling mechanism is the brain. (The brain is not itself involved in perception or thought, thinks Aristotle.) In humans, the area around the heart is particularly hot—because of the complexity of the perceptions and memories that a rational creature has. So the brain has to be particularly large. Therefore humans—in order to fulfill their role as rational animals—must have large brains.

There are two things to note about such an explanation. First, it is unashamedly teleological: that is, the parts of animals are to be explained according to their purpose or benefit, and, in particular, the purpose or benefit they bring to the function of the organism as a whole—in the case of a human, the function of thinking (i.e., being rational). Second, Aristotle seems not to have got things quite right. He is right that humans have large brains for an animal, but is it true that the heart is the organ of perception? That the brain plays no role in perception? That the brain cools the body down? Here we are confronted with some of Aristotle's failures: these are failures of observation, but they are failures that have consequences, because if you make a false move early on in science (already in the *Parts of Animals*, book II, Aristotle says that the brain plays no role in perception), the effects will be felt downstream (when you consider the brain's interaction with other organs, for instance).

Nonetheless, Aristotle's boundless energy in collecting observations and offering explanations is everywhere manifest. All his zoological works—and there are many of them, all presenting research that no one else had ever undertaken—are replete with observations about the animal kingdom, together with little explanations for why the animals in question are like they are. These explanations are waiting to be welded together into a scientific whole and presented as a unified theory (like Euclid's *Elements*), but because the animal kingdom is so vast, and because Aristotle is but one person, and because he was the first person, with no giants' shoulders to stand on, he did not get as far as that. The *Parts of Animals*, the *History of Animals*, the *Movement of Animals*, the *Generation of Animals*, treatises on sleep and dreams and memory (collected as the *Parva Naturalia*)—all of these works contain scientific observation and local explanations, all of which Aristotle hoped could be cast into syllogisms that would render perspicuous the different types of causes visible in nature, but all starting from the basic definitions of the various types of animal and their surrounding environments. Humanity is still seeking completeness in these matters two and a half thousand years later. One man alone—the inventor of the whole project—should not be expected to have completed that task, or even to have got right all of what he did do.

Ethics

Aristotle's life, dedicated to such extraordinary intellectual pursuits, must have been exhilarating. Even when confronted with the most ugly inhabitants of the animal kingdom, Aristotle enthusiastically points out the pleasure to be afforded by musing on the explanations of why those animals are made that way, and memorably quotes the story of Heraclitus who is supposed to have bidden strangers come into his humble kitchen, because "here too there are gods."[1] When he turned his mind to thinking about human life, and how best we can live it, he did not forget this joy. Of all the lives that one can lead, Aristotle singles out the one dedicated to intellectual pursuits, the one that gives the life of the mind pride of place, as the finest life one can lead—superior even to the life of the politician who is dedicated to a life of honorably serving their community. To understand why this is the case, we must once again turn to Aristotle's vision of what a human truly is.

Humans, you will recall, are rational creatures, but our rationality is a very distinctive one. Our rationality is dual. When we are doing science, and reflecting on the way things are and why they are that way, we are thinking about the finest and noblest things, the things that never change. To be sure, the natural world is a world teeming with change, but when we do logic and science we are looking for the unchanging patterns that explain and make sense of it—we are, in fact, thinking in the same way that gods do. Gods, for Aristotle, are the divine beings whose impassive thoughts are the drivers for the motions of the heavens above, a theory that Aristotle offers in his *Metaphysics*, in which he turns his attention upwards from the messy natural world around us to the perfect and untroubled world above. So that is one part of our rationality. But there is another, all too human, part: that which reflects on the questions of everyday life, where things are always changing, where there are not unchanging patterns enabling us to make sense of everything. These questions are the familiar ones of our day-to-day existence: What should I do with my life? Should I call my parents more often? What should I say to my friend who just failed to show up for a dinner we arranged? What should I say to my friend who just lost a loved one? Should I be giving more to charity than I do? Am I overindulging in the pleasures of food/drink/sex? Or, for that matter, am I underindulging in them?

Aristotle observes that humans have a distinctive kind of reason to deal with these pressing everyday matters, but also that there is another facet of our soul that engages with such matters: our emotions. We have bodily desires, such as hunger, thirst, lust, and cravings of all sorts; we also get scared of things, feel pity, care about what others think of us, have a conscience, get embarrassed, regret our actions, etc. These emotions are closely allied to our reasonings about daily life, and reflect the importance we attach to things: you (normally) don't feel shame in doing something if you think it's the right thing to do; you (normally) don't fear things if you don't think they are dangerous; you don't pity someone if they are leading a life you deem to be worthwhile. Our emotional reactions, in other words, answer to our reason. But there is another, remarkable, thing about emotions: we can change or shape our emotional reactions (Aristotle calls this *habituation*). For example, to encourage children to share their cookies, parents reward them when they do; that way, the initial pain they feel at not getting to eat the cookie they had their heart set on is softened; then next time, they share more easily, and the reward can be less. Over time, their adverse emotional reaction disappears, which opens the way for them to see what's good about sharing. It is also possible for adults to shape our own emotions. Perhaps you love donuts, but then read that online article about how bad donuts are for your health (at least in the quantities you are eating them). What to do? You might decide to cut down your donut consumption radically, but unfortunately, donut cravings don't disappear overnight. What you need to do is act on your decision this time (don't eat the donuts), and then on the next occasion make the same decision again and act on it. Gradually, over time, your desire for eating a dozen donuts at one sitting will subside. Desires that aren't indulged wither.

Humans are emotional but rational creatures; neither aspect could or should be eliminated, but what we do need to do is bring our emotions in line with our reason. Reason should set the agenda, but we should pay careful attention to our emotional well-being. After all, we often act with emotions as our guide—we often follow our heart. This is fine, as long as our heart is already in line with our reason. But what kind of reasoning do we use when we are reflecting on how to lead our life? In science, logic rules. In ethics, a different kind of thinking—more supple, more situation-dependent—needs to be in play. Aristotle calls this kind of thinking *deliberation*. He does not give many examples of this kind of thinking, but the following will give the flavor of what he has in mind: Suppose it is my partner's birthday, and I recognize that I should be doing something extra-special for her. I need to reflect on how best to ensure that she has a good time. Once I have settled on a course of action—perhaps taking her to her favorite restaurant—that action becomes for me a new goal for my deliberations. But my deliberations are not over yet: should I invite other people to join us? If so, who? Should I tell the kitchen it is her birthday, or would that embarrass her? Aristotle resolutely refuses to give rules for how to settle such questions, and it is important to note that in the domain of everyday life things are very unstable. Unlike in geometry, or even in natural science, there is no recourse to a set of fixed principles, the definitions and axioms that are characteristic of a science. After all, even the thing I am deliberating about—how to do something extra-special for my partner—is situation-specific: the week after her birthday, this will no longer be called for. And the way to create an extra-special occasion for her now may be to take her out to her favorite restaurant, but ten years ago she would have preferred to go out to her favorite bar. (The geometer does not have to keep up with the times like this!) So I have to know what she likes, and keep that knowledge appropriately up to date. Moreover, I have to ensure that when I am deliberating about what to do for her, I don't surreptitiously slip into thinking about what *I* would like to be doing for her birthday. It's her favorite restaurant that I have to call to mind, not mine.

Ethical deliberation is hard, and getting it right is very hard—harder, in some ways, than theoretical thinking, because of the huge numbers of variables involved. But Aristotle is optimistic that we can achieve the correct balance of rational discernment and deliberative expertise, coupled with emotional reactions that are in line with the actions that are correct: this balance he calls *virtue*. In the *Nicomachean Ethics*, he argues that we can't be happy as human beings without achieving that balance, since we are, by our natures, rational and emotional creatures. And the best way to achieve that balance is to have a life that is full of opportunity to engage in the distinctively human business of having family and friends and community, with philosophical activity to turn to when your obligations and commitments in those other areas are fulfilled.

Aristotle did not get everything right in ethics. When considering the nature of the best kind of community, he made the gross error of thinking that there were people around for whom slavery was their natural condition, and he similarly thought that women were intellectually inferior to men. These errors are errors of substance, about which one should never be dismissive; they had a terrible influence in the Middle Ages and beyond. In presenting Aristotle's ethical framework, one needs to show how to adjust and adapt his thinking so as to avoid his mistakes. But to judge from the reaction of philosophers these days, an ethical system built around Aristotelian virtue, with its distinctive take on moral character and moral knowledge, is still considered a live option for a true theory of how to lead one's life.

NOTE

1. *Parts of Animals*, book I, chapter 5.

parallelis et consequenter axis terrę sit axi totius mundi parallelus.

2. motus terrę ē annuu[s] quo mouet[ur] circa solem, sub linea cęlesti que ē ecliptica in medio Zodiaci in qua sunt 12. signia cęlestia ex quo fit, ut sin=gulis mensibus videat[ur] nobis sol signum unum percurrere, non sane, q[uia] sol moveat[ur], sed q[uia] ipsa terra unum signum suo motu percurrerit, ut patet, nam existente terra in puncto, h, sub libra.

Tradition

and the Legacy of Aristotle

BARBARA SCALVINI

What comes to mind when you hear the word *tradition*? Awe at the world's beauty and complexity? An exuberant, inventive mind? Confident command of reason and logic? Probably not. Yet the roots of *tradition* actually reach far from the connotations of quaintness, hidebound devotion to musty habits, or resistance to innovation that the term generally evokes today. From the Latin *tradere*, meaning "to hand down," the word more aptly celebrates the values, ideals, and ideas that have been bequeathed to us. It describes the rich inheritance that we cherish and are committed to nurturing in the next generation. Tradition and innovation are twins, the legacy of our cultural past, the foundation of our present, and a heritage that shapes our future. This perspective is key to understanding the legacy of Aristotle, and how his exceptional contributions have been handed down, with little or no interruption, for more than two millennia.

The philosophical journey of Aristotle's work is one of the richest, most astonishingly fecund, and most radically important intellectual adventures in the Western tradition. And it also is a compelling story. The narrative begins with an extraordinary man and is carried by countless people and episodes. It takes us from Athens to Syria, Baghdad, central Spain, Venice, Paris, Leipzig, and Oxford. It leads to the Europe-wide Renaissance, the "republic of letters," and then on to its central stage in Rome. There,

Cesare Cremonini (1550–1631)
Disputatio ultima; De coelo et mundo
Italy, c. 1630
(detail of cat. 25)

the landmark trial of Galileo, who turned against the prevailing ideas of Aristotle and other Greek philosophers, might have seemed to drive a nail into Aristotle's coffin. But no—"The Philosopher," as he had come to be known, sprung up as lively as ever. Embraced or rejected, imitated or misunderstood, Aristotle remained a formidable opponent: unavoidably confronted by scientists, an irreplaceable guide to navigating the pitfalls of knowledge-building in Enlightenment Germany and Scotland, an inspiration to America's founding fathers, and a fresh fountain for modern proponents of ethics based on virtue.

This exhibit and catalogue trace the several routes along which Aristotle's works traveled. The first is translations, which continually renewed Aristotle's work, presenting it through an ever-changing lens of linguistic and cultural contexts. Then come the commentaries, revealing the engrossing minds that have confronted this body of thought through the centuries. Finally come the books themselves, the material texts, the silent artifacts that physically convey words and thought passed from hand to hand—telling their stories through vellum and paper, printing type, manuscript, marginalia, page layouts, bindings, and notes of ownership.

Translations

In the early centuries of the common era, Aristotle's work was disseminated in books by his pupils: copies, translations, and commentaries. It did not command exceptional attention, and, with the exception of one or two Latin versions of the logical works, survived only among Eastern scholarly communities such as the Syrians, who were conquered by the Arabs around 650 CE. The Arab conquerors certainly paid attention to the wealth of knowledge preserved by the Syrians. In the flourishing House of Wisdom of ninth-century Baghdad, a community of scholars translated into Arabic the body of Aristotle's works that had been preserved, along with early Greek commentaries. Syriac and Arabic languages are structurally very different from Greek, so this endeavor necessarily became one of paraphrasing and interpretation—a fertile appropriation bringing a layer of newness. This did not just promote Islamic Aristotelianism, which continued robustly into the Renaissance (cat. 3), it protected the body of work from extinction and preserved it as a living tradition through Muslim conquests in Europe.

In Islamic Spain, rulers set out to gather books from across the Arab world. Many Arab scholars who had studied Greek ideas in the East introduced Aristotle's thought to the West. Notable among them was Ibn Rushd, known as Averroes, whom philosophers and theologians of the Middle Ages later called "The Commentator." By the twelfth century, much of Spain was no longer in Arab hands. Areas that had been conquered or inhabited by other civilizations over the centuries, enriched by linguistic perspectives from these varied cultures, became ideal settings for translation from Arabic into Latin. These translations were produced by very diverse scholars: Christians from Spain and Italy (Gerard of Cremona among the most productive), but also Jewish scholars translating from Hebrew, bringing their own interpretations that contributed to the richness and complexity of the work.

Soon the translation movement spread Europe-wide (cat. 13), with English and Scottish scholars participating in the rediscovery and diffusion of Aristotle's thought (cat. 31). This movement ushered in the great age of humanism. Knowledge of Greek, the rediscovery of Greek manuscript sources, and a new spirit of meticulous faithfulness to original sources brought about landmark editions, such as Theodorus Gaza's exceptionally widespread translation of the natural science sections of the work (cat. 16), and that of his "competitor," George of Trebizond. By the early sixteenth century, Aristotle's work was solidly and pervasively established throughout Europe in Latin, which remained the universal language of learning up to the end of the eighteenth century and beyond.

Commentaries

While it may be an exaggeration to claim that the entire history of philosophy is a commentary on Plato and Aristotle, the fact remains that this centuries-long conversation was the prime mode of philosophical reflection for many generations. Philosophers regarded Aristotle's texts as an authority to be organized (in digestible partitions: e.g., book, chapter, question), explained (by exegesis or minute exposition), made relevant (paraphrased), summarized (in epitomes), and critically discussed (in glosses).

Editing at this level involved a full and complex commitment both to the text itself and to the particular readership expected to benefit from the commentary. Alexander of Aphrodisias (cat. 2), one of the earliest scholars to hold a chair of Aristotelian philosophy by imperial appointment, does not provide a diligent exposition of Aristotle's thought on the soul and on destiny so much as an original investigation of the human soul based on principles established by Aristotle, reflecting the concerns of a new age. Eustratius, the twelfth-century metropolitan bishop of Nicaea, made his exceptionally influential commentary a wholly original Neoplatonic interpretation of Aristotle's *Nicomachean Ethics* (cat. 36). This was then adopted by theologians such as Albert the Great and Bonaventure. Thomas Aquinas incorporated Aristotle's logical method; proof of a "prime mover"; views of place, time, and motion; views of sensory perception and intellect; and much more in his *Summa Theologiae*, perhaps the most complete articulation of medieval Christian philosophy (cat. 33).

The Martin J. Gross Collection, which has gathered all these works together, comes into its own when it supplements printed editions of commentaries with a largely underexplored and diverse variety of lecture notes and handwritten annotations. Lectures by the Jesuit Francisco Suárez, which survive only in a few manuscripts, transform Aristotle's logic into a powerful, refined tool (cat. 8). The handwritten annotations of a student of Scottish philosopher James Martin accompanying a printed edition of Aristotle, as yet unpublished, stand among the very few witnesses to Aristotle's critical views on the eternity of the world (cat. 30). A previously unknown Spanish commentary reveals an interpretation of Aristotle's logic separate and different from the contemporary and well-known reading of the Coimbra school (cat. 9). The Latin version of the Greek humanist John Argyropoulos is elaborated upon in print by two Leipzig academics of the early sixteenth century, and further developed by another in manuscript notes, perhaps in preparation for a new edition (cat. 35). And the lecture notes of Cesare Cremonini (cat. 21) reveal how the robustness of Aristotle's system "forced" Galileo to provide a theoretical

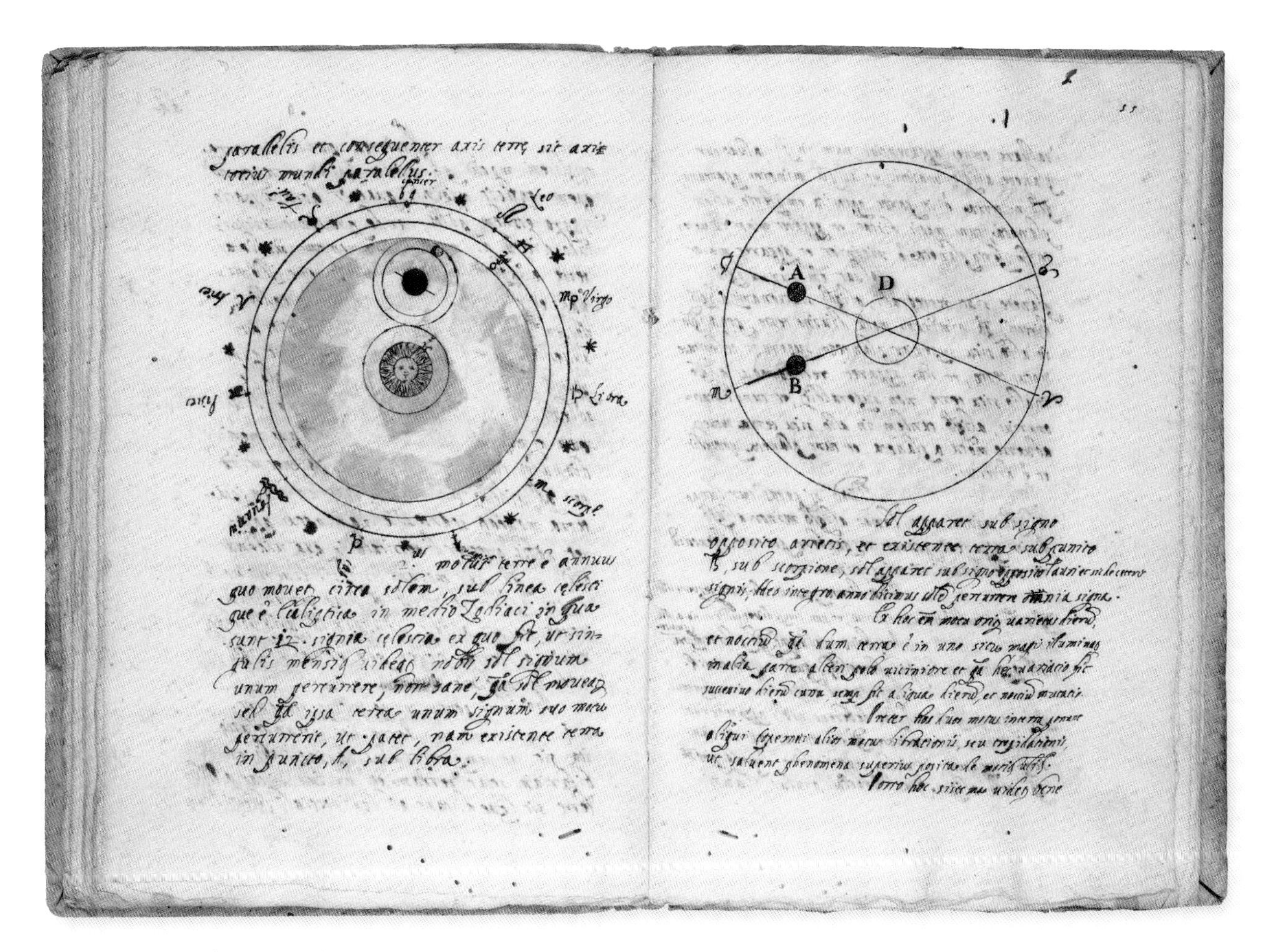

Cesare Cremonini (1550–1631)
Disputatio ultima; De coelo et mundo
Italy, c. 1630 (cat. 25); a problem in optics (left)

outlook to the ballast of his observations. Each volume opens a door to a new chapter in the history of thought.

Material Texts

As we approach these books, we are confronted not only with abstract ideas but with tangible artifacts that speak to us as eloquently as the words they contain. We may not ordinarily pause to consider that it takes one whole sheep to obtain a single largish sheet of parchment, and a fifty-strong herd simply to enable some sage's notions on the world to be written down—or indeed, some *other* sage's commentary on such notions.

Each compendium takes hundreds of hours with quills and ink, and hundreds of expensive candles to work by in the winter. Yet people did it, with the dedication and assurance of those who take up a mission for humanity for generations to come. They filled parchment sheets with minute letters thick with abbreviations. They exploited every space with economy and clarity by subtly allotting different locations or markers to text, gloss, and notes in ways that even now are almost instantly understandable to readers (cat. 13).

The advent of the printing press opened a world of new opportunities. A single edition, produced perhaps in the thousands, could have provided identical copies to readers throughout Europe. This, in turn, could "fix" or normalize one particular version over others. Aldus Manutius—a genial man who understood the significance of the printing-press revolution and drove its success with an exceptional series of editions—was the first publisher with the courage, skills, means, and vision to produce an edition of the entire known Aristotelian corpus. His landmark publication is still considered one of the most important books ever published (cat. 1).

Although the texts in printed editions were comparatively standardized, individual copies still retained a wealth of individual features. The manuscript annotations in the copy of the Aldus edition account for an enormous portion of the content, minutely reflecting the thought of at least three Renaissance readers. A copy of a Paris edition of Aristotle's text on colors, with the printed translation and commentary of Simone Porzio (a student of the philosopher Pietro Pomponazzi), has in its margins the annotations of Guillaume Chrestien, the physician to King Francis I and King Henry II of France (cat. 20). Francesco Buonamici, Galileo's teacher, discusses Aristotle's *Posterior Analytics* and its ancient interpreter Philoponus, using that as a framework for the "new science" (cat. 5) in a dialogue that strings together a millennium's worth of thought with a creative—indeed, revolutionary—outcome.

A visually striking manuscript examining the *Physics* is redacted in microscript—tiny, barely visible letters of ever decreasing size (cat. 29). This perhaps suggests that the author felt it necessary to be cryptic in order to protect himself from unfriendly eyes. Openly challenging a cosmology with Earth at its center would not have been acceptable in all circles. Yet, one scholar at least could not resist the impulse to find out the truth and record it, by whatever means necessary.

And then there are the bindings. A 1606 manuscript commentary on logic, in remarkable Spanish calf with gilt motifs, points tantalizingly to a connection with contemporary Jewish circles (cat. 9). Armorial decorations on a commentary on natural philosophy, which often mentions Descartes, suggest that it may have been produced in a Protestant region of France (cat. 26). The manuscript of the unpublished lectures of the Jesuit Giuseppe Agostini, prepared for the Roman College and richly bound for a Roman patrician, show, in their sympathy for Galilean positions, that the philosophical and scientific conversation was not confined to the academy (cat. 22).

The story of Aristotle's philosophy has continued into modernity. It might be tempting to relegate him firmly to a remarkable but dead past. After all, major advancements in science have made his specific observations on the natural world, and some of his views on society, quite obsolete: we now know about gravity, thermodynamics, and atoms; we no longer see Earth as the center of the universe, nor do we see slavery as an acceptable component of a well-functioning community. Yet the dialogue between a fast-changing world and this profound philosopher has continued to be as creative as it ever was. Seventeenth-century thinkers took Aristotle's logic to a new level of complexity. Thus, for Leibniz, logic became the manipulation of symbols according to established rules, a sort of operation that can be freed from error or subjectivity and performed one day by machines. Modern computers, which have changed the way we do most things since World War II, are our contemporary realization of this approach. The founding fathers of the United States were keen to embrace and reinterpret Aristotle's belief, which still resonates today, that governments should rule for the good of the people, not for the good of those in power, and that a "pure democracy" can easily turn into a tyranny of the majority. Aristotle's idea of teleology—that is, the notion that all things tend toward an end or goal—has informed fields as diverse as medical ethics and economics. Each book and manuscript featured here bears witness to an ongoing dialogue with the Aristotelian corpus. They underscore the differing ways in which tradition can be harnessed in support of change and how the past is always present.

Aristotil
mus fe

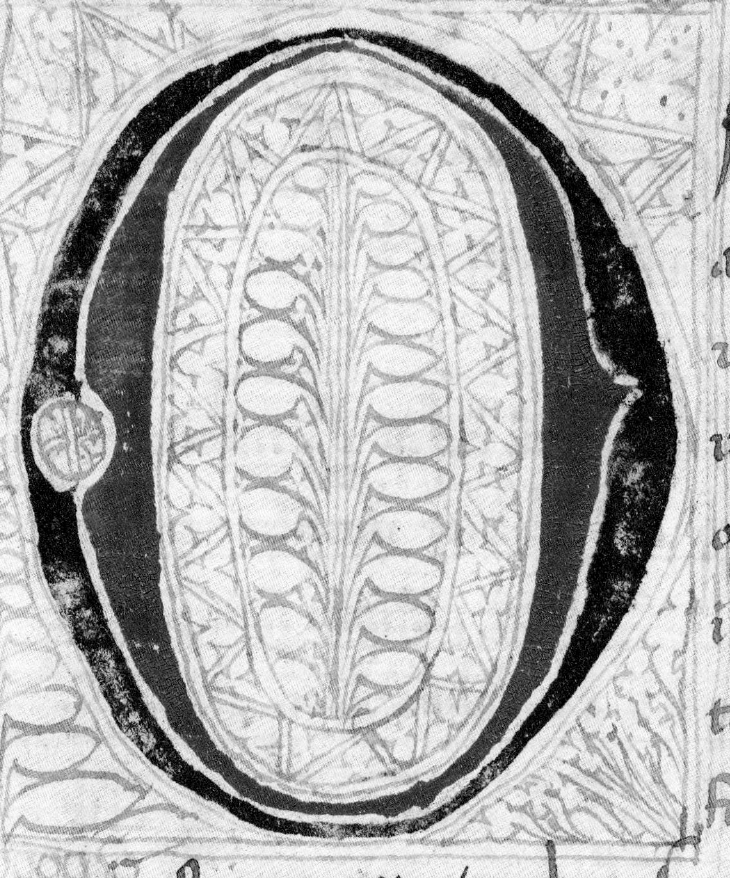

mnis doctria & c
actus bonu qddaz appe
tur bonu qd oia app
um/ii qd e sut opatio
quedaz. Quor aut
i hiis meliora existut
tioibus, entibus, & arti
fines. Medicinalis
no est navigatio. militaris aut e victoria. yconom
sut taliu sub una qdaz virtute quemadmodu sub equestri
istoru sut. h at & ois bellica opatio sub militari. s3
In oibus itaq3 architectoicaru fines oibus sut desidera
e. oia enim h persequntur. differt at nichil operationes ipas ee fines ac
quemadmodu in dictis doctrinis. Si utiq3 aliqs finis e opabiliu q
ipm & no oia pp alteru desideramus. procedit enim in infinitu
siderium. Manifestum qm hic utiq3 erit bonu & optimu e.
habet in cognitione quemadmodu sagittatores signu habentes
Si aut sic teptandu est figuraliter accipere ipm qd e & cuius disc
libus

CATALOGUE

1

THE ALDUS ARISTOTLE

"What an impetus the learning of the world would receive if the treasures of Greece and Rome were made available to a great number by being in printed form."[1]

ALDUS MANUTIUS

Aristotle

Opera [5 volumes, in Greek]

Venice: Aldus Manutius, 1495–98

Aldus Manutius was born in 1450, at a time when Gutenberg was experimenting with printing from moveable type. A classical scholar by training, Manutius learned Latin studying in both Rome and Ferrara before turning his attention to Greek. He met and studied with Greek scholars who had fled Constantinople after its invasion by the Ottoman Turks, bringing with them many Greek manuscripts previously unknown in Europe. In 1482 Manutius became a tutor to the sons of the Prince of Carpi, and it was then that he resolved to merge his knowledge of the ancient authors with his desire to print books containing the classical literature of the ancient world.

By the last decade of the fifteenth century, only a few books in Greek had been printed, and only one of them was substantial—the 1488 two-volume edition of Homer's works printed in Florence by Bernardus and Nerius Nerlius, with types cut by Demetrius Damilas. In 1490 Manutius wrote:

> I have resolved to devote my life to the cause of scholarship. I have chosen in place of a life of ease and freedom, an anxious and toilsome career. A man has higher responsibilities than the seeking of his own enjoyment; he should devote himself to an honorable labor. Living that is a mere existence can be left to men who are content to be animals. Cato compared human existence to iron. When nothing is done with it, it rusts. It is only through constant activity that polish or brilliancy is secured.[2]

Manutius began his journey by printing the Greek and Latin grammar of Constantine Laskaris (1435–1501) in 1494–95. The second, and perhaps greatest, achievement of his press was the publication between 1495 and 1498 of the known works of Aristotle—their first appearance in Greek, a genuine monument in the history of ideas and the history of printing and publishing. This was the beginning of Manutius's extraordinary effort to print the entire corpus of Greek and Latin classics.

Aldus Manutius,
19th century,
wood engraving
after contemporary
copper engraving.
Alamy Photo

Printing the Aristotle was a huge undertaking in terms of the sheer volume of text, the technicalities of printing, and the cost. It was a landmark in humanist scholarship, craftsmanship, and ingenuity. The Greek fonts, which had to be produced in great quantities, with separate characters for accentuation, were based on the calligraphic hand of a leading Greek scribe. They were cut by Francesco Griffo, who also designed the famous Aldine italic—still a favorite font today.

These particular five volumes—sold separately and at different prices—have been together in a complete set for more than five hundred years. They are also exceptional for the amount and complexity of the annotations left by early humanist readers. There are at least three overlapping sets of commentary, each revealing a distinctive set of responses to Aristotle, evidence of how seriously Aristotle was read in the early modern period. At the turn of the sixteenth century, his works remained a living presence in the intellectual culture of the day. It was indeed due to the efforts of Aldus Manutius that "the light of Greek letters returned to shine on civilized peoples."[3]

NOTES

1. Quoted in William Dana Orcutt, *The Kingdom of Books* (Boston: Little, Brown & Co., 1927), 7.
2. Quoted in George Haven Putnam, *Books and Their Makers During the Middle Ages*, vol. 1, 476–1500 (London: G. P. Putnam's Sons, 1898), 418.
3. Translation of a plaque on the house of Aldus Manutius in Venice.

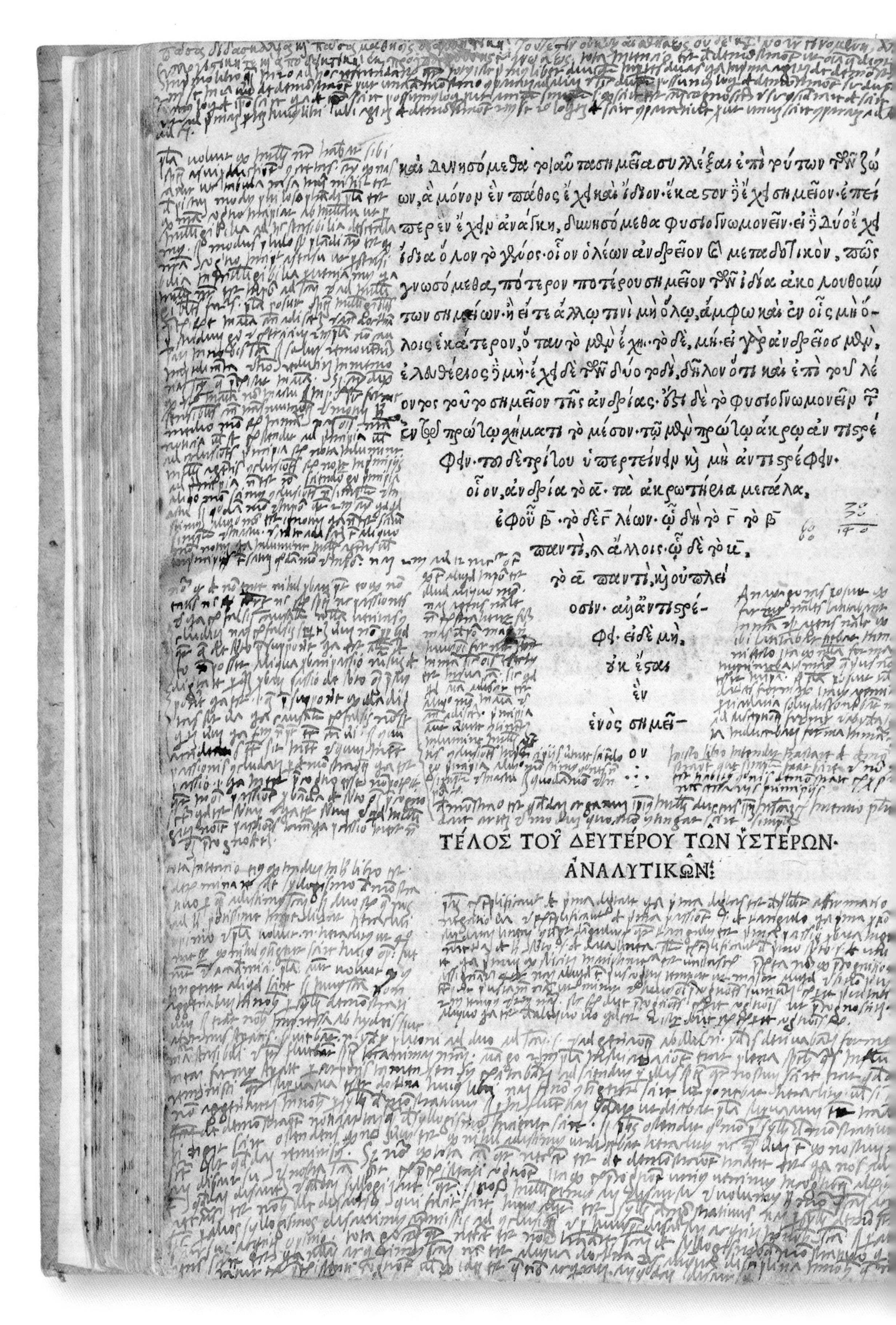

καὶ δυνησόμεθα τοιαῦτα σημεῖα συλλέξαι ἐπὶ τούτων τῶν ζώων, ἃ μόνον ἓν πάθος ἔχει ἴδιον· ἕκαστον δ' ἔχει σημεῖον· ἐπείπερ ἓν ἔχειν ἀνάγκη, δυνησόμεθα φυσιογνωμονεῖν. εἰ δὲ δύο ἔχει ἴδια ὅλον τὸ γένος, οἷον ὁ λέων ἀνδρεῖον καὶ μεταδοτικόν, πῶς γνωσόμεθα, πότερον ποτέρου σημεῖον τῶν ἰδίᾳ ἀκολουθούντων σημείων; ἢ εἴ τε ἄλλῳ τινὶ μὴ ὅλῳ, ἄμφω καὶ ἐν οἷς μὴ ὅλοις ἑκάτερον, ὅταν τὸ μὲν ἔχῃ, τὸ δὲ μή. εἰ γὰρ ἀνδρεῖος μέν, ἐλευθέριος δὲ μή, ἔχει δὲ τῶν δύο τοδί, δῆλον ὅτι καὶ ἐπὶ τοῦ λέοντος τοῦτο σημεῖον τῆς ἀνδρείας. ἔστι δὲ τὸ φυσιογνωμονεῖν τῷ ἐν τῷ πρώτῳ σχήματι τὸ μέσον, τῷ μὲν πρώτῳ ἄκρῳ ἀντιστρέφειν, τοῦ δὲ τρίτου ὑπερτείνειν καὶ μὴ ἀντιστρέφειν. οἷον, ἀνδρεία τὸ α. τὰ ἀκρωτήρια μεγάλα, ἐφ' οὗ β. τὸ δὲ γ λέων. ᾧ δὴ τὸ γ, τὸ β παντί, ἀλλὰ καὶ ἄλλοις. ᾧ δὲ τὸ β, τὸ α παντί, καὶ οὐ πλείοσιν, ἀλλ' ἀντιστρέφει. εἰ δὲ μή, οὐκ ἔσται ἓν ἑνὸς σημεῖον.

ΤΕΛΟΣ ΤΟΥ ΔΕΥΤΕΡΟΥ ΤΩΝ ὙΣΤΕΡΩΝ ἈΝΑΛΥΤΙΚΩΝ.

Cat. 1. End of *Prior Analytics* and beginning of *Posterior Analytics*, with extensive annotations

ΑΡΙΣΤΟΤΕΛΟΥΣ ΑΝΑΛΥΤΙΚΩΝ ΥΣΤΕΡΩΝ ΗΤΟΙ ΤΗΣ ΑΠΟΔΕΙΚΤΙΚΗΣ ΠΡΩΤΟΝ.

ΠΑΣΑ διδασκαλία καὶ πᾶσα μάθησις διανοητικὴ ἐκ προϋπαρχούσης γίνεται γνώσεως. φανερὸν δὲ τοῦτο θεωροῦσιν ἐπὶ πασῶν. αἵ τε γὰρ μαθηματικαὶ τῶν ἐπιστημῶν διὰ τούτου τοῦ τρόπου περαίνονται καὶ τῶν ἄλλων ἑκάστη τεχνῶν. ὁμοίως δὲ καὶ περὶ τοὺς λόγους, οἵ τε διὰ συλλογισμῶν καὶ οἱ δι' ἐπαγωγῆς. ἀμφότεροι γὰρ διὰ προγινωσκομένων ποιοῦνται τὴν διδασκαλίαν. οἱ μὲν λαμβάνοντες ὡς παρὰ ξυνιέντων, οἱ δὲ δεικνῦντες τὸ καθόλου διὰ τοῦ δῆλον εἶναι τὸ καθ' ἕκαστον. ὡσαύτως καὶ οἱ ῥητορικοὶ συμπείθουσιν. ἢ γὰρ διὰ παραδειγμάτων, ὅ ἐστιν ἐπαγωγή, ἢ δι' ἐνθυμημάτων, ὅπερ ἐστὶ συλλογισμός. διχῶς δ' ἀναγκαῖον προγινώσκειν. τὰ μὲν γὰρ ὅτι ἔστι, προϋπολαμβάνειν ἀναγκαῖον. τὰ δὲ τί τὸ λεγόμενόν ἐστι, ξυνιέναι δεῖ. τὰ δ' ἄμφω. οἷον ὅτι μὲν ἅπαν ἢ φῆσαι ἢ ἀποφῆσαι ἀληθές, ὅτι ἔστι. τὸ δὲ τρίγωνον, ὅτι τόδε σημαίνει. τὴν δὲ μονάδα ἄμφω, καὶ τί σημαίνει καὶ ὅτι ἔστιν. οὐ γὰρ ὁμοίως τούτων ἕκαστον δῆλον ἡμῖν. ἔστι δὲ γνωρίζειν τὰ μὲν πρότερον γνωρίζοντα, τῶν δὲ καὶ ἅμα λαμβάνοντα τὴν γνῶσιν, οἷον ὅσα τυγχάνει ὄντα ὑπὸ τὸ καθόλου, ὧν ἔχει τὴν γνῶσιν. ὅτι μὲν γὰρ πᾶν τρίγωνον ἔχει δυσὶν ὀρθαῖς ἴσας, προῄδει. ὅτι δὲ τόδε τὸ ἐν τῷ ἡμικυκλίῳ τρίγωνόν ἐστιν, ἅμα ἐπαγόμενος ἐγνώρισεν. ἐνίων γὰρ τοῦτον τὸν τρόπον ἡ μάθησίς ἐστι, καὶ οὐ διὰ τοῦ μέσου τὸ ἔσχατον γνωρίζεται, ὅσα ἤδη τῶν

Cat. 1. *Topics*

ΑΡΙΣΤΟΤΕΛΟΥΣ ΤΟΠΙΚΩΝ ΠΡΩΤΟΝ.

Ἡ ΜΕΝ Πρόθεσις τῆς πραγματείας, μέθοδον εὑρεῖν, ἀφ' ἧς δυνησόμεθα συλλογίζεσθαι περὶ παντὸς τοῦ προτεθέντος προβλήματος ἐξ ἐνδόξων· καὶ αὐτοὶ λόγον ὑπέχοντες, μηθὲν ἐροῦμεν ὑπεναντίον. πρῶτον οὖν ῥητέον, τί ἐστι συλλογισμὸς· καὶ τίνες αὐτοῦ διαφοραί, ὅπως ληφθῇ ὁ διαλεκτικὸς συλλογισμός. τοῦτον γὰρ ζητοῦμεν κατὰ τὴν προκειμένην πραγματείαν. ἔστι δὴ συλλογισμὸς, λόγος ἐν ᾧ τεθέντων τινῶν, ἕτερόν τι τῶν κειμένων ἐξ ἀνάγκης συμβαίνει, διὰ τῶν κειμένων. ἀπόδειξις μὲν οὖν ἐστιν, ὅταν ἐξ ἀληθῶν καὶ πρώτων ὁ συλλογισμὸς ᾖ, ἢ ἐκ τοιούτων, ἃ διά τινων πρώτων καὶ ἀληθῶν τῆς περὶ αὐτὰ γνώσεως τὴν ἀρχὴν εἴληφε. διαλεκτικὸς δὲ συλλογισμός, ὁ ἐξ ἐνδόξων συλλογιζόμενος. ἔστι δὲ ἀληθῆ μὲν καὶ πρῶτα, τὰ μὴ δι' ἑτέρων ἀλλὰ δι' αὑτῶν ἔχοντα τὴν πίστιν. οὐ δεῖ γὰρ ἐν ταῖς ἐπιστημονικαῖς ἀρχαῖς ἐπιζητεῖσθαι τὸ διὰ τί· ἀλλ' ἑκάστην τῶν ἀρχῶν, αὐτὴν καθ' ἑαυτὴν εἶναι πιστήν. ἔνδοξα δὲ τὰ δοκοῦντα πᾶσιν, ἢ τοῖς πλείστοις, ἢ τοῖς σοφοῖς· καὶ τούτοις, ἢ τοῖς πᾶσιν, ἢ τοῖς πλείστοις, ἢ τοῖς μάλιστα γνωρίμοις καὶ ἐνδόξοις. ἐριστικὸς δέ ἐστι συλλογισμός, ὁ ἐκ φαινομένων ἐνδόξων, μὴ ὄντων δέ· καὶ ὁ ἐξ ἐνδόξων δέ, ἢ φαινομένων ἐνδόξων φαινόμενος. οὐ γὰρ πᾶν τὸ φαινόμενον ἔνδοξον, καὶ ἔστιν ἔνδοξον. οὐδὲν γὰρ τῶν λεγομένων ἐνδόξων ἐπιπόλαιον ἔχει παντελῶς τὴν φαντασίαν, καθάπερ περὶ τὰς τῶν ἐριστικῶν λόγων ἀρχὰς συμβέβη-

ARISTOTELIS STAGIRITAE

OMNIA, QVAE EXTANT, OPERA,

NVNC PRIMVM SELECTIS TRANSLATIONIBVS,
EMENDATIONIBVS EX COLLATIONE
græcorum exemplarium, ſcholiis in margine illuſtrata, nouo
etiam ordine digeſta: Additis præterea non nullis
libris nunquam antea latinitate donatis.

AVERROIS CORDVBENSIS IN EA OPERA
Omnes, qui ad nos peruenere, Commentarii.

NON NVLLA SVPER ADDITA DVBIA, FIGVRAE,
notationes, nunquam antea editæ, vt Auerrois media in libros Metaphyſ.
commentatio: eiuſdem de Spermate libellus.

GRAECORVM, ARABVM, ET LATINORVM MONV-
menta quædam, ad hoc opus ſpectantia.

MARCI ANTONII ZIMARAE IN ARIST. ET AVER. DICTA CONTRA-
dictionum ſolutiones, quibus nunc addidimus doctiſsimorum
virorum ſolutiones 100.

HAEC AVTEM OMNIA TVM EX PRAEFATIONE,
tum ex indice librorum clarius innoteſcunt.

Cum ſummi Pontificis, Gallorum Regis, Senatusq; Veneti decretis.

VENETIIS MDLX.

LOGIC AND METAPHYSICS

"All humans by nature desire to know."

METAPHYSICS, BOOK I, CHAPTER 1

WHAT IS THE SUBSTANCE OF A THING? Can a sentence be both true and false, or can two contradictory statements be true at the same time, in the same sense? Can we establish relationships of cause and effect between things or events? Is every actual thing or event preceded by a state of "being potential"? How can we ensure that our reasoning is a universally acceptable string of arguments? We may think these questions empty and obvious, but if we do so it is simply because we have relied on logical methods so thoroughly that they have become hidden to us. After all, Alfred North Whitehead and Bertrand Russell employed many pages of their *Principia Mathematica* (1910–13) to prove that 1 + 1 = 2.

The way Aristotle investigated the world, trying to reveal its hidden structures, has defined the field of philosophy. The texts featured here—from the ancient commentators Alexander of Aphrodisias and Johannes Philoponus to obscure teachers of the Renaissance, from Spanish and Portuguese professional logicians to early modern philosophers confronted with the first telescopes—give us a view into how we have harnessed the perception of our senses and the basic rules of logic to build a large body of knowledge and actively transform the world.

Aristotle and Averroes (Ibn Rushd) (1126–1198)
Aristotelis ... opera ... Averrois Cordubensis in ea opera omnes ... commentarii
Venice: Comin de Trino di Monferrato, 1560
(detail of cat. 3)

2

PRESERVING ARISTOTLE

Alexander Aphrodisaeus (2nd–3rd century CE)

De anima ex Aristotelis institutione

Brescia: Bernardinus de Misintis, 1495

Aristotle lived and taught in Greece in the fourth century BCE. After his death in 322 BCE, and the deaths of his first-generation pupils, the preservation and study of his works fell into disarray and neglect. The philosopher Alexander of Aphrodisias, or Aphrodisaeus, a Peripatetic—meaning a follower of the Aristotelian school—rose to meet the challenge. His extensive commentaries on Aristotle, notable for their precision and exactitude, shaped the tradition of Aristotelian texts for centuries. This book, for example, contains the first appearance in print of his commentary on Aristotle's *On the Soul*. Its controversial contention is that the individual reasoning faculty is inseparable from the body, and that therefore the individual soul is mortal. Despite ecclesiastical condemnation in the Middle Ages, Alexander's theory thrived, fostered by the esteem in which his works were held by the Arabs, who produced translations and commentaries from the ninth to the twelfth century. Alexander's idea of the individual mortality of the soul shaped the philosophy of some of the major thinkers of the Renaissance, most notably Pietro Pomponazzi (1462–1525) and Simone Porzio (1496–1554; cat. 20). Pomponazzi's works were controversial in his own time, but Porzio did not experience any backlash, and the concept survived through the nineteenth century and beyond.

HIERONYMI DONATI PATRICII VE/
NETI IN INTERPRETATIONEM ALE/
XANDRI APHRODISEI.

PRAEFATIO.

Riſtotelem philoſophum ita ab initio antiquitas ad mirata eſt: ut eum lauda/ ret potius q̃ ſectaretur. Nam & in græcia & in ita lia diu Platonem ſequuti ſunt: ambitione & pompa maiore q̃ ſtudio. Quod & carminibus & dialogis & affectatis tranſlationi bus rerum naturæ traditione propoſita delectabã tur. Quam docendi rationem: iam per ea tempora receptam: quom: uelut a philoſophiæ grauitate & ſimplicitate alienam omnino repudiaſſet Ariſtote les: difficilior & preſſior iiſdem temporibus uiſus eſt. Sed cum in diſponenda doctrinarum inſtitu/ tione mire doctus & atticus. tum in inceſſendis antiquis philoſophis aſſidue & fœliciter pertinax. philoſophie tandem apud poſteros principatũ op/ tinuit: ut multis iam ſæculis longe plures ſub Ari/ ſtotele q̃ ſub Platone profeciſſe uideantur. Non

a 2

Cat. 2. Preface

3

THE ARAB CONTRIBUTION

Aristotle and Averroes (Ibn Rushd) (commentator, 1126–1198)

Aristotelis ... opera ... Averrois Cordubensis in ea opera omnes ... commentarii

Venice: Comin de Trino di Monferrato, 1560

This landmark edition of Aristotle's works includes the extensive commentaries by the Arab philosopher Ibn Rushd, whose name was Latinized as Averroes, a major figure in the transmission of Aristotle's thought. Arab scholarship was fundamental in preserving Aristotle's works. After the initial interest of an early generation of thinkers, such as Alexander of Aphrodisias (cat. 2), Aristotle attracted little attention except from Syrian copyists. When the Arabs conquered Syria in the seventh century CE, they inherited and cherished those texts. It was in ninth-century Baghdad, in the House of Wisdom, that Aristotle's works and the earliest commentaries were translated into Arabic. This was a complex and creative cultural synthesis, since Syriac and Arabic are structurally quite different from Greek. Thus Aristotle continued to live in Islamic philosophy; it was through Islam that Aristotle reached the West, following Arab conquests in Spain.

At the magnificent scholarly courts of Islamic Spain, rulers patronized Arab scholars who were well versed in Greek philosophy. Among them was the towering figure of Averroes, whom philosophers and theologians of the Middle Ages later called "The Commentator." At caliph Abu Ya'qub Yusuf's behest, and continuing during his successor's reign, Averroes wrote a series of summaries and commentaries on most of Aristotle's works, all of which contributed immensely to his contemporaries' understanding of Aristotle and the early Greek and Arab commentators, such as Themistius, Alexander of Aphrodisias, al-Farabi, and Avicenna (Ibn Sina). Averroes's works influenced Jewish and Christian thought in the following centuries. His original Arabic writings survived from the thirteenth century, chiefly thanks to translations into Hebrew conducted by Jews, who, driven out of southern Spain, took them north to France. It was from the Hebrew versions that a fresh program of Latin translations flourished in the Renaissance, culminating with this complete, accurate edition, published (again) in Venice.

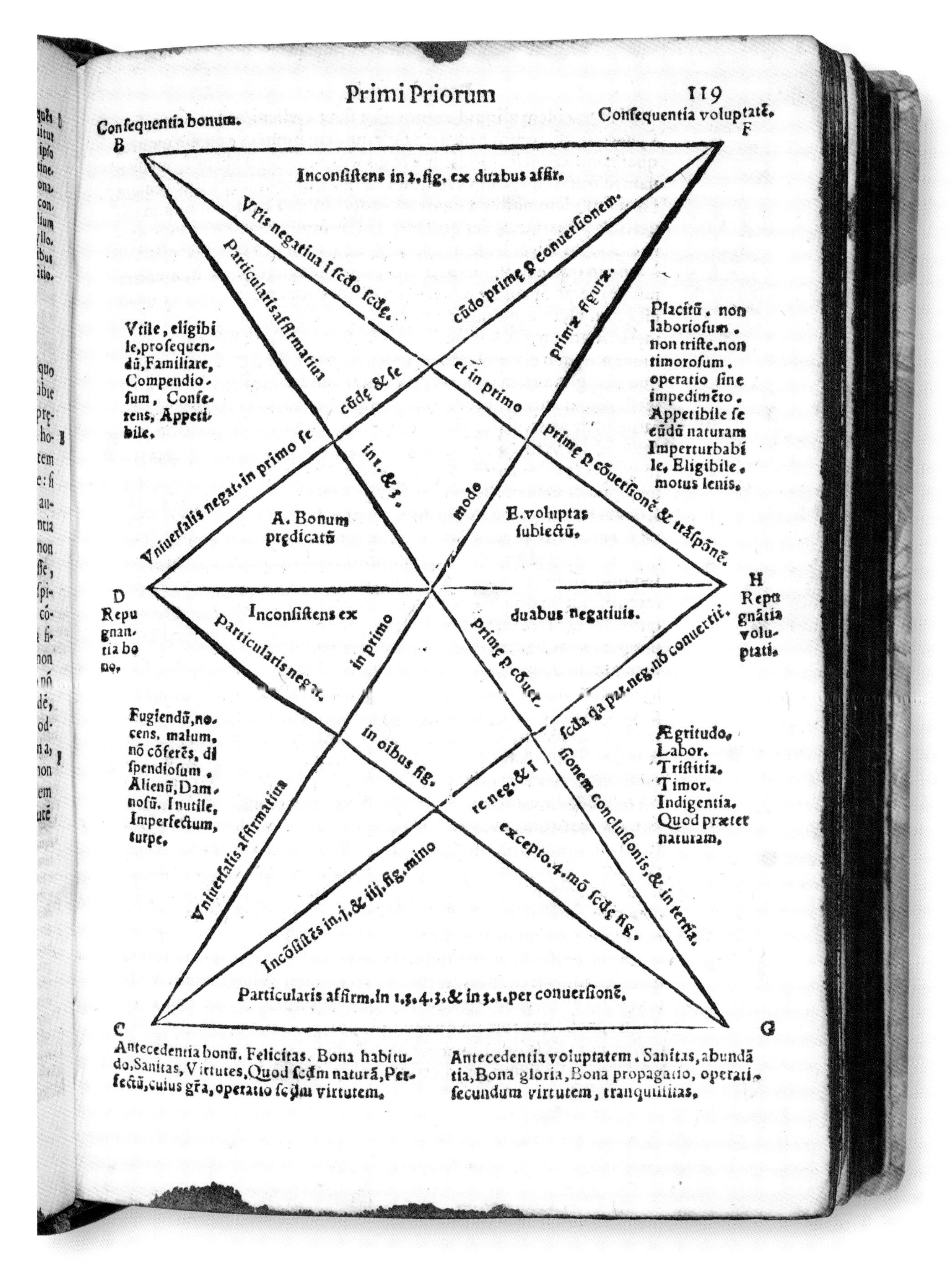

Cat. 3. Categories and relationships in ethics: a diagram visualizing the interplay between what is good in itself and what is delectable or desirable

4
THE METAPHYSICS

Aristotle

Duodecim libri Metaphisice

Leipzig: Martin Landsberg, 1503

The word *metaphysics* today refers to the branch of philosophy concerned with the nature of reality and being, but originally it simply meant "what is placed after the *Physics*" in the sequence of Aristotle's works and in school curricula. Aristotle in his *Metaphysics*, having described the physical world, goes on to talk about the law of non-contradiction, by which two contradictory statements cannot be true at the same time in the same sense; actual and potential states of being; the nature of substance; and the nature of causality. These concepts are still debated today.

This edition of the *Metaphysics* was printed and bound in Germany, in the university city of Leipzig. This copy gives us a fascinating insight into Renaissance book production. At the time, books were not sold as we know them—that is, bound and priced. Buyers would have gone to a printer or bookseller and seen rows and rows of blocks of printed sheets neatly stacked; they would have chosen a work, bought the sheets after negotiating a price, and then taken the sheets to a trusted binder. They would have chosen a style for the cover, and, as in the case of the owner of this book, might have instructed the binder to insert blank leaves, alternating with the printed ones. This practice, called *interleaving*, together with the addition of blanks bound at the beginning or at the end of a volume, was favored by scholars, who used the additional leaves to record their observations on the text.

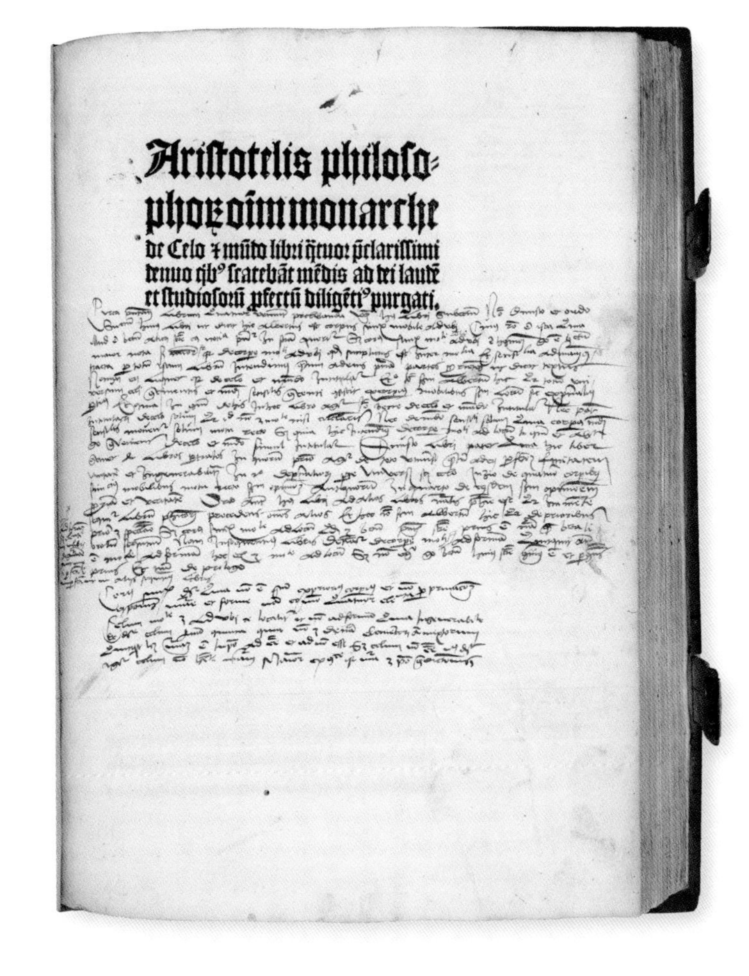
Aristotelis philoso-
phoꝝ oīm monarche
de Celo ⁊ mūdo libri q̄tuor p̄clarissimi
denuo q̇b⁹ scatebāt mēdis ad dei laudē
et studiosorū p̄fectū diligēti⁹ purgati.

Cat. 4. *On the Heavens*

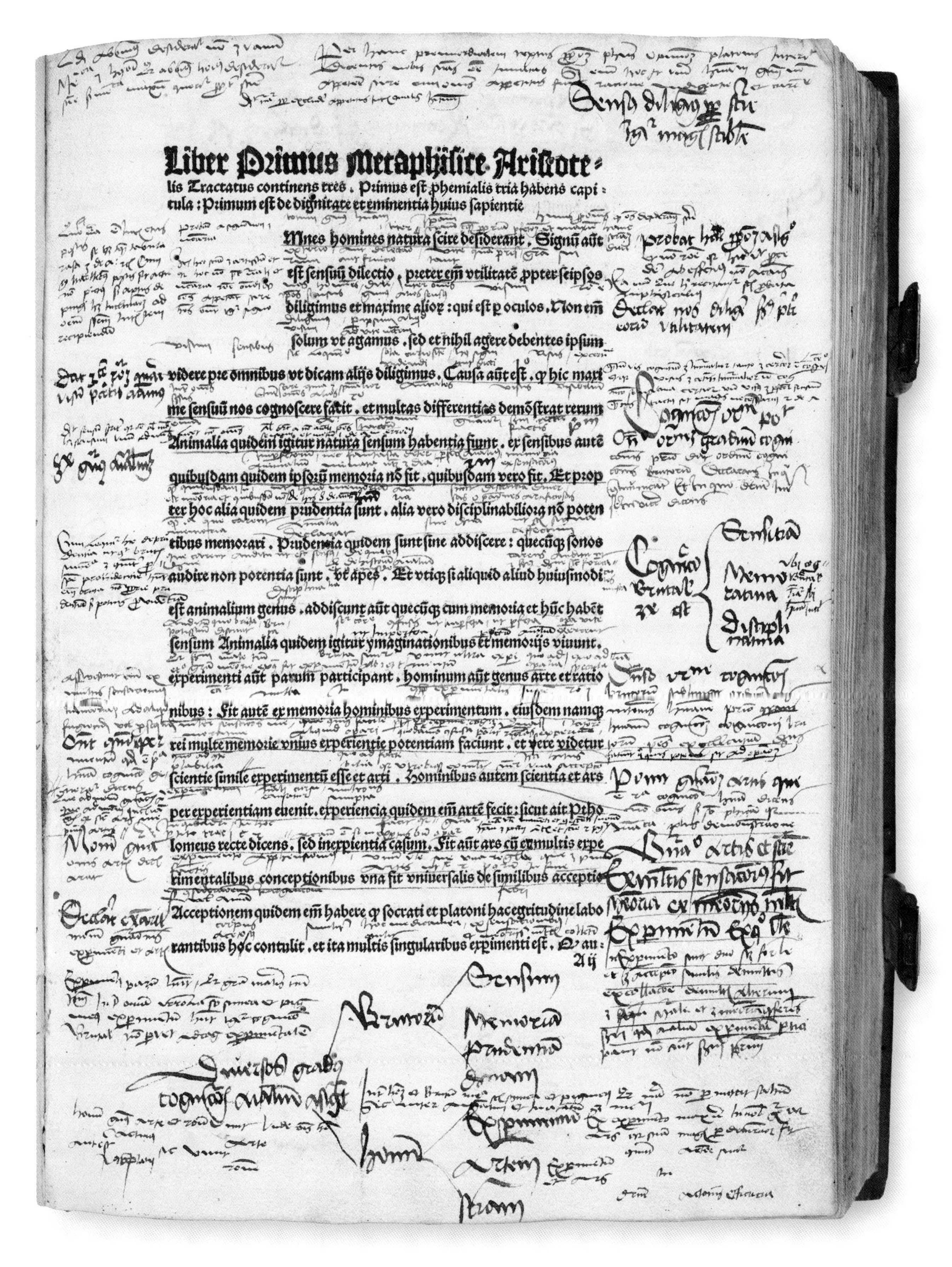

Liber Primus Metaphisice Aristote-
lis Tractatus continens tres. Primus est phemialis tria habens capi-
tula: Primum est de dignitate et eminentia huius sapientie

Omnes homines natura scire desiderant. Signū aūt
est sensuū dilectio, preter enim vtilitatē ppter seipsos
diligimus et maxime alio℞: qui est p oculos. Non enim
solum vt agamus, sed et nihil agere debentes ipsum
videre pre omnibus vt dicam alijs diligimus. Causa aūt est, q hic maxi
me sensuū nos cognoscere facit, et multas differentias demōstrat rerum
Animalia quidem igitur natura sensum habentia sunt, ex sensibus autē
quibusdam quidem ipso℞ memoria nō fit, quibusdam vero fit. Et prop
ter hoc alia quidem prudentia sunt, alia vero disciplinabilio℞ nō poten
tibus memorari. Prudentia quidem sunt sine addiscere: quecūq sonos
audire non potentia sunt, vt apes. Et vtiq si aliquid aliud huiusmodi
est animalium genus, addiscunt aūt quecūq cum memoria et hūc habēt
sensum Animalia quidem igitur ymaginationibus et memorijs viuunt,
experimenti aūt parum participant, hominum aūt genus arte et ratio
nibus: Fit autē ex memoria hominibus experimentum, eiusdem namq
rei multe memorie vnius experientie potentiam faciunt, et vere videtur
scientie simile experimentū esse et arti. Hominibus autem scientia et ars
per experientiam euenit, experiencia quidem enim artē fecit: sicut ait Ptho
lomeus recte dicens, sed inexpientia casum. Fit aūt ars cū ex multis expe
rimentalibus conceptionibus vna fit vniuersalis de similibus acceptio
Acceptionem quidem enim habere q socrati et platoni hac egritudine labo
rantibus hoc contulit, et ita multis singularibus experimenti est. Q au:
A ij

Cat. 4. The *Metaphysics*, with annotations visualizing the roles of senses, memory, and rational faculties in the cognitive process

5

GALILEO'S TEACHER

Johannes Philoponus (5th–6th century CE)

In Posteriora ... Aristotelis commentaria

Venice: Manutius, 1534

Annotated by Francesco Buonamici (1533–1603)

We are all familiar with the phrase *eureka moment*, that thrilling instant in which a new, surprising understanding comes to us like a bolt of lightning. In the case of the transmission of Aristotle, we can talk about a eureka stretch, when a wholly new understanding instead spreads gradually and inexorably. This book is a snapshot of just such a eureka stretch: it was owned by Francesco Buonamici (1533–1603), and it shows traces of a new science dawning in the mind of a remarkable teacher, who had as a student at the University of Pisa the young Galileo Galilei (1564–1642).

This book is a Venetian edition of a commentary on Aristotle's *Posterior Analytics* by Johannes Philoponus. In the *Posterior Analytics*, Aristotle moves from examining logic in its internal consistency to looking at its consistency with the workings of the material world. Although the text attracted several commentaries, that of Philoponus was popular for centuries. Philoponus was a forward-looking philosopher of the fifth–sixth century CE who used the genre of commentary to expound his own theories—in this case, the notion of impetus in describing movement. The publishers of this work were the heirs of the great Aldus Manutius (1449–1515), and they brought to their work the same scholarly discipline and focus that Manutius himself did (see cat. 1). They were well aware that Aristotle's text had inspired groundbreaking works of science in antiquity, and that it would be beneficial to contemporary scientists to read such works from readily available

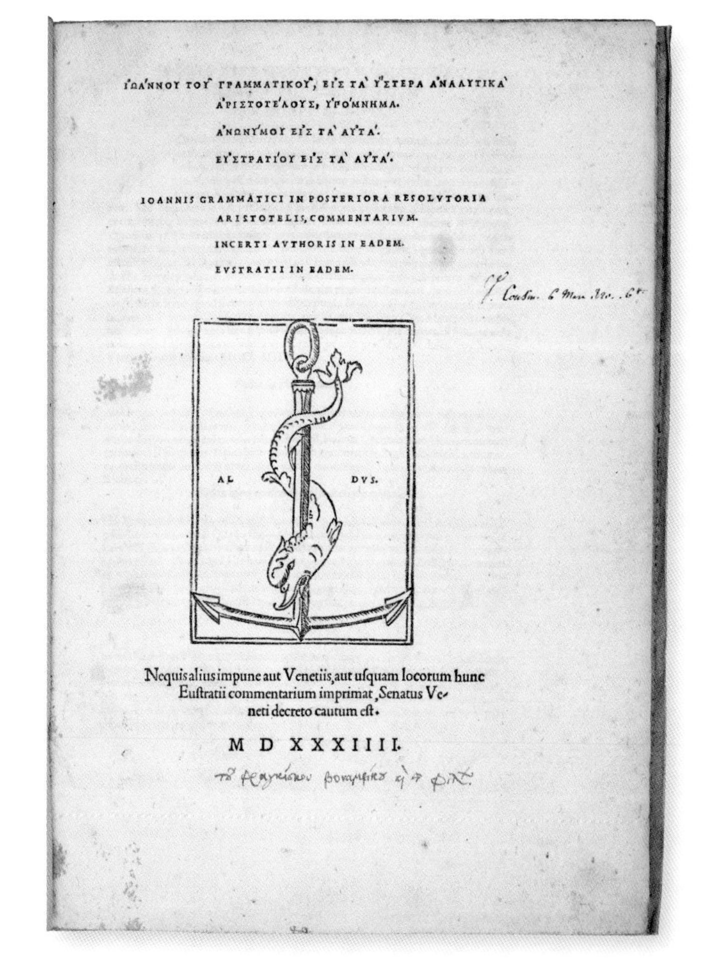

ἸΩΆΝΝΟΥ ΤΟΥ͂ ΓΡΑΜΜΑΤΙΚΟΥ͂, ΕἸΣ ΤᾺ ὙΣΤΕΡΑ ἈΝΑΛΥΤΙΚᾺ
ἈΡΙΣΤΟΤΈΛΟΥΣ, ὙΠΌΜΝΗΜΑ.
ἈΝΩΝΎΜΟΥ ΕἸΣ ΤᾺ ΑὐΤΆ.
ΕὐΣΤΡΑΤΊΟΥ ΕἸΣ ΤᾺ ΑὐΤΆ.

IOANNIS GRAMMATICI IN POSTERIORA RESOLVTORIA
ARISTOTELIS, COMMENTARIVM.
INCERTI AVTHORIS IN EADEM.
EVSTRATII IN EADEM.

Ne quis alius impune aut Venetiis, aut usquam locorum hunc Eustratii commentarium imprimat, Senatus Veneti decreto cautum est.

M D XXXIIII.

Cat. 5. Title page with ownership inscription of Buonamici beneath the date

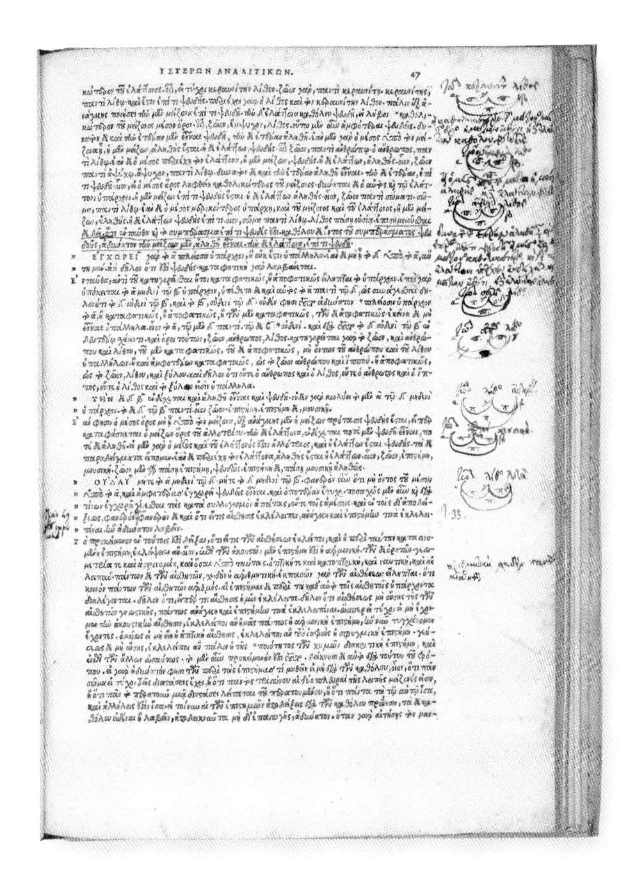
ΥΣΤΕΡΩΝ ΑΝΑΛΥΤΙΚΩΝ. 47

ΦΙΛΟΠΟΝΟΥ ΕΙΣ ΤΟ ΠΡΩΤΟΝ.

Cat. 5. Marginal annotations showing principles and demonstrations of syllogisms, penned by Galileo's teacher, Francesco Buonamici

printed editions rather than from mostly inaccessible, fragmentary, and often error-ridden manuscripts.

Instructive as the printed text is in itself, the margins of this copy are filled with hundreds of Buonamici's notes and diagrams on space, place, time, and matter—the beginnings of a whole new chapter in the history of science. These notes may reflect the lectures that Galileo heard when he enrolled at Pisa in 1580, at the age of sixteen. Later, Galileo annotated his own copy of the same commentary, demonstrating its importance in the development of his thought, particularly concerning his discovery of the law of uniformly accelerated motion.

A missing link between Aristotle's physics and the nascent new science of the time comes to life before our eyes as we leaf through this book. It allows us to understand Galileo's frames of reference, the reasons why he set out his arguments the way he did, the source of the terms he used, and the roots of the problems he set out to solve. This book gives us a visual measure of the value of tradition in the handing down of a text from one philosopher-scientist to another. It reminds us that often, even in science, new insights are born out of a history, and answers are found according to how the question has been framed.

6

A MOST UNORTHODOX PRIEST

Johannes Philoponus (5th–6th century CE)

Commentaria in libros De anima Aristotelis

Venice: Bartholomeo Zanetti, 1535
Annotated by Pandolfo Ricasoli (1581–1657)

This is the first appearance in print of the influential commentary on Aristotle's *On the Soul* by Johannes Philoponus. Philoponus was a theologian as well as a philosopher, and his works stand at the beginning of a long Christian engagement with Aristotle. This engagement can be seen here in his reflections on the nature of sensory perception, the role of images in thought, and the immortality of the soul. His commentary on *On the Soul* is perhaps the earliest in which he begins to criticize Aristotelian doctrine rather than just comment upon it. In 681 CE, roughly a century after his death, Philoponus was censured as a heretic by the third Council of Constantinople because of his ideas on the Trinity.

In fact, this copy comes from the library of a heretical priest, Pandolfo Ricasoli (1581–1657), who was condemned for putting into practice his own theories on the soul, involving a considerable commitment to sex. The book is extensively annotated in Greek, likely by Ricasoli himself, and the notes are interspersed with an alphanumerical key, which could indicate that he kept his commentary in a separate notebook to allow himself more room to write. Ricasoli joined the Florentine Theological College in 1611 and became a canon in 1620. He is known for holding the remarkable belief that the soul is raised to God through sexual intercourse. Not only are carnal acts permissible, they are to be positively encouraged. Sexual relations allow the body to be free from ordinary sensory perception, while the soul is elevated to the divine. Ricasoli enthusiastically put this theory into practice at the Casa di Santa Dorotea, a school for girls in Florence. In his capacity as spiritual director of the school, Ricasoli applied himself to a thorough exploration of his doctrine, before being apprehended by the Florentine Inquisition (to whom he was already known as the owner of a considerable collection of prohibited books). He was excommunicated, and sentenced to prison for life in November 1641. He died in the prisons of Santa Croce in 1657.

ΙΩΑΝΝΟΥ ΑΛΕΞΑΝΔΡΕΩΣ, ΤΟΥ
φιλοπόνου εἰς τὸ περὶ ψυχῆς Ἀριστοτέλους σχο-
λικαὶ ἀποσημειώσεις ἐκ τῶν συνουσιῶν
Ἀμμωνίου τοῦ Ἑρμείου μετά τινων
ἰδίων ἐπιστάσεων.

ΠΡΟΟΙΜΙΟΝ.

ΕΛΛΟΝΤΑΣ ἡμᾶς τῶν περὶ ψυχῆς ἀκροᾶσθαι λόγων, ἀναγκαῖον εἰπεῖν πρότερον περὶ τῶν δυνάμεων τῆς ψυχῆς, ποσαχῶς τε διαιροῦνται, καὶ ποίας ἑκάστη τῆς ὀνομασίας τετύχηκεν. εἶτα πόσαι τῶν
ἀρχαιοτέρων περὶ αὐτῶν δόξαι γεγόνασι. καὶ ἐπὶ τούτοις τὴν ἀληθῆ περὶ
αὐτῶν δόξαν ἐκ διαιρέσεως ἀφορίσασθαι. πρῶτον μὲν αἱ ψυχικαὶ δυνάμεις, τὴν εἰς δύο διαίρεσιν ἐπιδέχονται. αἱ μὲν γὰρ αὐτῶν εἰσὶ λογικαί.
αἱ δὲ, ἄλογοι. ἑκατέρα δὲ τούτων τῶν δυνάμεων διχῆ πάλιν διαιρεῖται.
τῶν γὰρ λογικῶν, αἱ μέν εἰσι ζωτικαὶ καὶ ὀρεκτικαί. αἱ δὲ, γνωστικαί. ὁ-
μοίως διαιροῦνται καὶ αἱ ἄλογοι. πάλιν δὲ, αἱ λογικαὶ καὶ γνωστικαὶ τῆς
ψυχῆς δυνάμεις διαιροῦνται τριχῆ. ἡ μὲν γάρ τις ἐστὶ δόξα. ἡ δὲ διάνοια.
ἡ δὲ νοῦς. Ἡ μὲν οὖν δόξα καταγίνεται περὶ τὸ καθόλου τὸ ἐν τοῖς
αἰσθητοῖς. τοῦτο γὰρ γινώσκει. οἶδε γὰρ ὅτι πᾶν λευκὸν, διακριτικὸν ὄψεως.
καὶ ὅτι πᾶς ἄνθρωπος, δίπους. ἔτι γε μὴν καὶ τῶν διανοητῶν, οἶδεν ἄνευ λόγου τὰ συμπεράσματα. οἶδε
γὰρ ὅτι ἡ ψυχὴ ἀθάνατος. διατί δὲ ἀθάνατος ἡ ψυχὴ, οὐκ οἶδεν ἡ δόξα, ὅτι διανοίας ἔργον ἐστὶ τοῦτο. δό-
ξης δὲ τὸ ὅτι μόνον εἰδέναι. ὥστε δόξα ἐστὶ τό, τε γνωστικὸν τὸ ἐν τοῖς αἰσθητοῖς καθόλου, καὶ τῶν διανοητῶν τὰ συμπεράσματα. ὅθεν καὶ καλῶς ὁρίζεται τὴν δόξαν ἐν τῷ σοφιστῇ διαλόγῳ ὁ ἐλεάτης ξένος,
διανοίας, λέγων [illegible] γὰρ διανοίας [illegible] ὅτι ἀθάνατος ἡ ψυχὴ, λαβοῦσα ἡ δόξα τὸ συμπέρασμα τοῦτο μόνον οἶδεν ὅτι ἀθάνατος. Διάνοια δέ ἐστι, τὸ οἷον ὁδόν τινα διανύειν, μεταβαί-
νουσαν ἀπὸ προτάσεων ἐπὶ συμπεράσματα, ἐξ οὗ καὶ τὴν κλῆσιν εἵλκυσεν. οἷον, πόθεν ὅτι ἡ ψυχὴ ἀθάνατος ζητεῖ ἡ διάνοια. εἶτα ἐκ τῶν σαφεστέρων ἀρξαμένη διαβαίνει ἐπὶ τὸ ζητούμενον. λέγουσα ὅτι ἡ ψυχὴ αὐτοκίνητος. τὸ αὐτοκίνητον καὶ ἀεικίνητον. τοῦτο δὲ ἀθάνατον. ἡ ψυχὴ ἄρα ἀθάνατος. καὶ τῆς μὲν διανοίας ἔργον τοῦτο. Τοῦ δὲ νοῦ ἔργον τὸ ἁπλαῖς προσβολαῖς καὶ κρεῖττον, ἢ κατὰ ἀπόδειξιν ἐπιβάλλειν
τοῖς πράγμασιν. ὥσπερ γὰρ ἡ αἴσθησις προσβάλλουσα τυχὸν τῷ λευκῷ, ἢ τῷδε τῷ σχήματι, κρεῖττον ἢ
κατὰ ἀπόδειξιν αὐτοῦ τὴν γνῶσιν ἔχει. οὐ γὰρ δεῖται πρὸς ταῦτα συλλογισμοῦ. ὅτι τὸ δέ ἐστι λευκόν. ἀλλὰ ἁπλῇ ἐπιβολῇ τοῦτο γινώσκει. οὕτω καὶ ὁ νοῦς ἁπλῇ ἐπιβολῇ γινώσκει τὰ νοητὰ, κρεῖττον ἢ κατὰ
ἀπόδειξιν. Ἡ δέ γε τοῦ νοῦ ἐνέργεια ἐκείνοις μόνοις παραγίνεται, οἳ εἰς ἄκρον καθάρσεως, καὶ ἐπιστήμης γεγόνεν ἀφικέσθαι, καὶ διὰ τῶν καθαρτικῶν ἀρετῶν ἀφαντάστως καὶ δίχα αἰσθήσεως ἐνεργεῖν
συνειθισμένοις. ἐστι γὰρ ὁ νοῦς οἷον ἕξις τῆς ψυχῆς τελειοτάτη. ὅθεν περὶ τούτου λέγων ὁ πλωτῖνός φησιν ὅτι
ὅστις δὲ ἐνήργησεν, οὗτος οἶδεν ὃ λέγω. ἅτε δὴ τῆς τοιαύτης καταστάσεως μηδὲ λόγῳ ἑρμηνευθῆναι δυ-
ναμένης. καταγίνεται τοίνυν ὁ μὲν νοῦς, περὶ τὰ νοητά. ἅπερ ὁ τίμαιος ὁριζόμενος ἔλεγεν. εἶναι νοήσει
μετὰ λόγου περιληπτά. ἡ δὲ διάνοια, περὶ τὰ διανοητά. ἡ δὲ δόξα, περὶ τὰ δοξαστά, ἅπερ καὶ αὐτὰ τὰ δο-
ξαστὰ λέγων ὁ τίμαιός φησιν. εἶναι δόξῃ μετ᾽ αἰσθήσεως ἀλόγου δοξαστά. τούτων δὲ τῶν δυνάμεων, πρώ-
την μὲν, ἐπέχει τάξιν ὁ νοῦς. ἐσχάτην δὲ, ἡ δόξα. μέσην δὲ, ἡ διάνοια. ἥτις καὶ προσῳκείωται τῇ ἡμετέρᾳ
ψυχῇ, ἐπειδὴ καὶ αὐτὴ τὴν μέσην τὴν ἐν τῷ παντὶ τάξιν ἔχει. καὶ διὰ ταύτης, λέγω δὲ τῆς διανοίας ἀνά-
γεται ἡ ἡμετέρα ψυχὴ ἐπὶ τὴν τῶν νοητῶν θεωρίαν. ἥτις ἐστὶ τελειότης τῆς ψυχῆς. ἐπειδὴ γὰρ σύντροφός
ἐστι καὶ σύμφυλος τοῖς αἰσθητοῖς ἡ ψυχὴ ἡ ἡμετέρα, ἀδυνατεῖ διὰ τὸν συνεθισμὸν τῶν αἰσθήσεων, ἐπὶ
τὴν θεωρίαν τῶν νοητῶν καὶ ἀΰλων ἀνάγειν ἑαυτήν. ἀλλὰ νομίζει κἀκεῖνα σώματα εἶναι, καὶ μεγέθη
» ἔχειν, καὶ ὅσα ἐπὶ τῶν αἰσθητῶν κἀκεῖ φαντάζεται. ὡς καὶ ὁ πλάτων ἐν τῷ φαίδωνι λέγει, ὅτι τοῦτό
» ἐστι χαλεπώτατον τῶν ἐν ἡμῖν, ὅτι ὅταν καὶ σχολὴν ἀπὸ τῶν περιολκῶν τοῦ σώματος μικρὸν ἀγάγω-
» μεν, καὶ θελήσωμεν τῇ θεωρίᾳ τῶν θείων σχολάσαι, παρεμπίπτουσα ἡ φαντασία θόρυβον ἡμῖν κινεῖ, ὑπο-

A

τῶν ψυχικῶν δυνάμεων διαίρεσις.

λογικαὶ πρῶτο.

δόξα τί.

Ζήνων ἐλεάτης.

διάνοια.

νοῦς.

πλωτῖνος.

πλάτων.

Cat. 6. Prologue

7

A SCHOLAR READS THE METAPHYSICS

Aristotle

Eorum quae Physica sequuntur, sive Metaphysicorum, ut vocant, libri tredecim

Paris: Thomas Richardus, 1564
Annotated by an unidentified scholar (16th century)

This copy of an edition of the *Metaphysics* allows us to look over the shoulder of a sixteenth-century scholar. Through the many annotations made by the unidentified scholar who owned the book, we can observe the teaching methods and techniques that an academic would have used in Renaissance Europe. The scholar begins his notes with a date: 20 October 1566, toward the start of the academic year. Leafing through the book we can almost see him as he structures the course and jots down lecture prompts. He adds many annotations to the chapters dealing with foundational concepts, the basic building blocks of reasoning that students must first be taught. Then he concentrates on more sophisticated points, such as causation, sensory perception, and memory. When it comes to the complex concept of matter, one quick lesson is not enough: he needs an entire extra leaf to discuss it. Can matter be divided ad infinitum? Or are we bound to find at some point small indivisible particles (in Greek, *atomos*)? Over half a millennium, countless teachers would have similarly recorded notes and observations as they prepared their classes.

Cat. 7.
Annotations

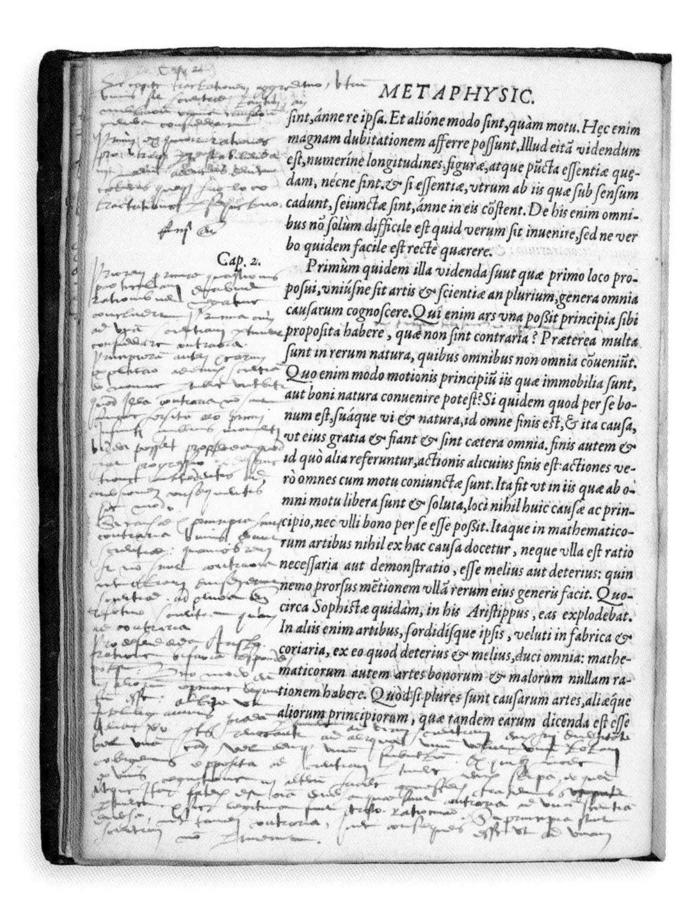

METAPHYSIC.

sint, ánne re ipsa. Et alióne modo sint, quàm motu. Hęc enim magnam dubitationem afferre possunt. Illud eitã videndum est, numeríne longitudines, figuræ, atque pũcta essentiæ quędam, nécne sint. & si essentiæ, vtrum ab iis quæ sub sensum cadunt, seiunctæ sint, ánne in eis cõstent. De his enim omnibus nõ solùm difficile est quid verum sit inuenire, sed ne verbo quidem facile est rectè quærere.

Cap. 2.

Primùm quidem illa videnda sunt quæ primo loco proposui, vniúsne sit artis & scientiæ an plurium, genera omnia causarum cognoscere. Qui enim ars vna poßit principia sibi proposita habere, quæ non sint contraria? Præterea multa sunt in rerum natura, quibus omnibus non omnia cõueniũt. Quo enim modo motionis principiũ iis quæ immobilia sunt, aut boni natura conuenire potest? Si quidem quod per se bonum est, suáque vi & natura, id omne finis est, & ita causa, vt eius gratia & fiant & sint cætera omnia. finis autem & id quò alia referuntur, actionis alicuius finis est: actiones verò omnes cum motu coniunctæ sunt. Ita fit vt in iis quæ ab omni motu libera sunt & soluta, loci nihil huic causæ ac principio, nec vlli bono per se esse poßit. Itaque in mathematicorum artibus nihil ex hac causa docetur, neque vlla est ratio necessaria aut demonstratio, esse melius aut deterius: quin nemo prorsus mẽtionem vllã rerum eius generis facit. Quocirca Sophistæ quidam, in his Aristippus, eas explodebat. In aliis enim artibus, sordidísque ipsis, veluti in fabrica & coriaria, ex eo quod deterius & melius, duci omnia: mathematicorum autem artes bonorum & malorum nullam rationem habere. Quòd si plures sunt causarum artes, aliæque aliorum principiorum, quæ tandem earum dicenda est esse

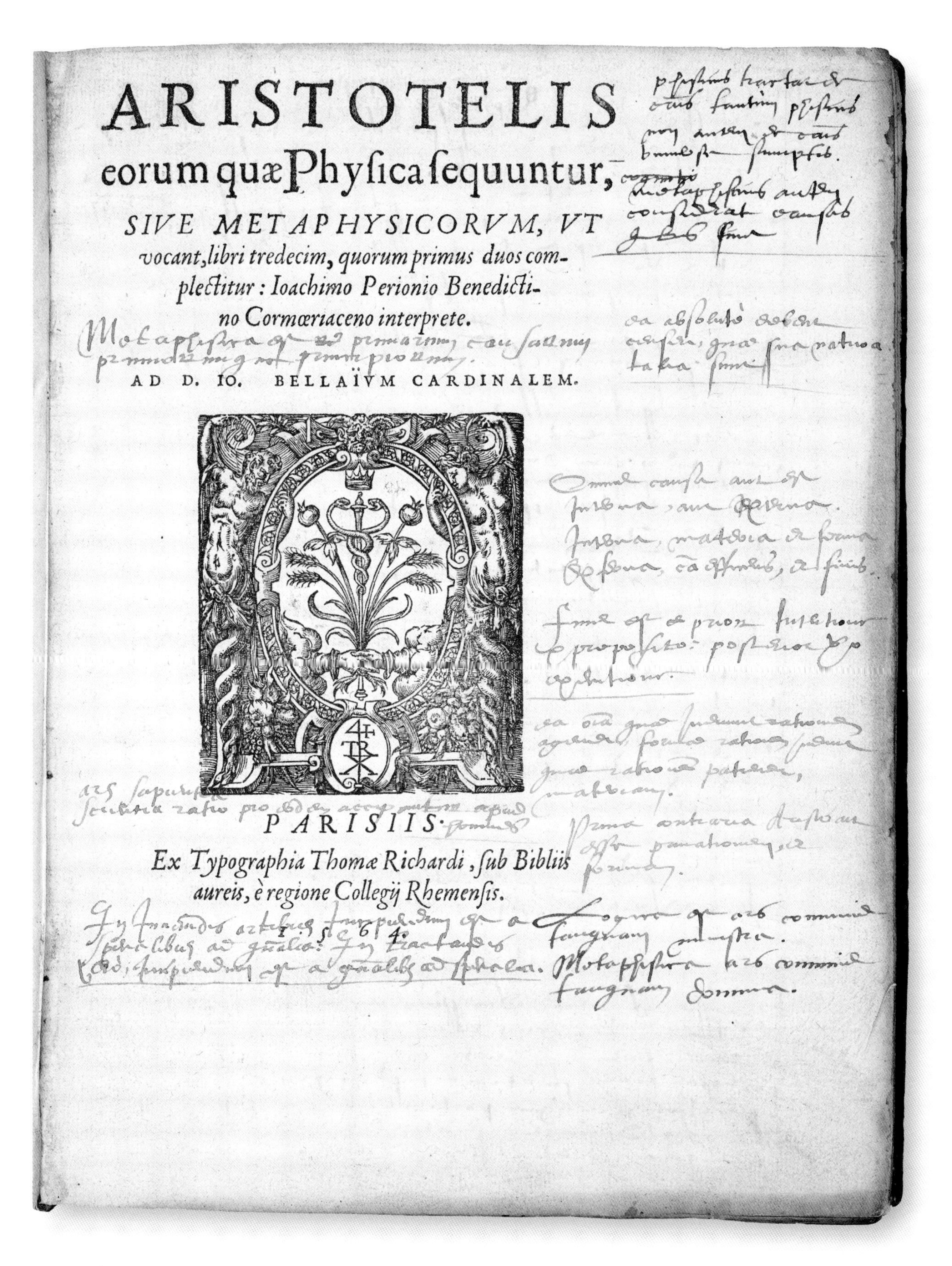

ARISTOTELIS
eorum quæ Physica sequuntur,
SIVE METAPHYSICORVM, VT
vocant, libri tredecim, quorum primus duos complectitur: Ioachimo Perionio Benedictino Cormœriaceno interprete.

AD D. IO. BELLAÏVM CARDINALEM.

PARISIIS.
Ex Typographia Thomæ Richardi, sub Bibliis aureis, è regione Collegij Rhemensis.
1564.

Cat. 7. Title page

8

THE JESUIT CONTRIBUTION

Francisco Suárez (1548–1617)

[**Commentary on Aristotle's logical works**]

Rome, 1583
Manuscript on paper

Francisco Suárez hailed from a *converso* or Marrano family—that is, Jews who converted to Christianity rather than being expelled from Spain in the late fifteenth century. He is arguably one of the most important figures in the transmission of medieval philosophy in the early modern era. His key work, *Disputationes Metaphysicae*, which appeared in Salamanca in 1597, consisted of fifty-four questions dealing with metaphysics. Suárez's career is interwoven with the activity of the Jesuits, who conducted serious scholarly activities in Europe beginning in the sixteenth century.

The Roman College (Collegio Romano) was the intellectual powerhouse for Jesuits. Given the college's proximity to the papal court and the Jesuits' special relationship with the papacy, it could be assumed that the college would have been deeply conservative. On the contrary, it was home to remarkably progressive thinkers, and Francisco Suárez was one of the most illustrious. He was a prolific writer and profound thinker whose company and counsel were sought by popes and kings alike. His writings, both books and lectures, took Aristotle's works and made them powerful tools for handling the challenges of the early modern world. Starting from Aristotle, he developed original positions on the relationship between church and state, on logic, on metaphysics, and notably on international law, a discipline that counts him as one of its founders. His belief that any political order has human, not divine, origins, and that a prince can therefore be legitimately deposed by the people, was strongly contested in his own day and afterward.

These manuscript lecture notes are penned in a neat, fair handwriting, likely by a student or a clerk. They are not what is called a "working copy," full of small, compressed notes that only the writer himself could easily decipher; instead they appear to have been prepared for public or semipublic reading, for a person of regard who might have requested them, or possibly as a template for a publication. Their content is different from any of Suárez's works printed at the time, and they are in need of further study to assess their impact on his later thinking.

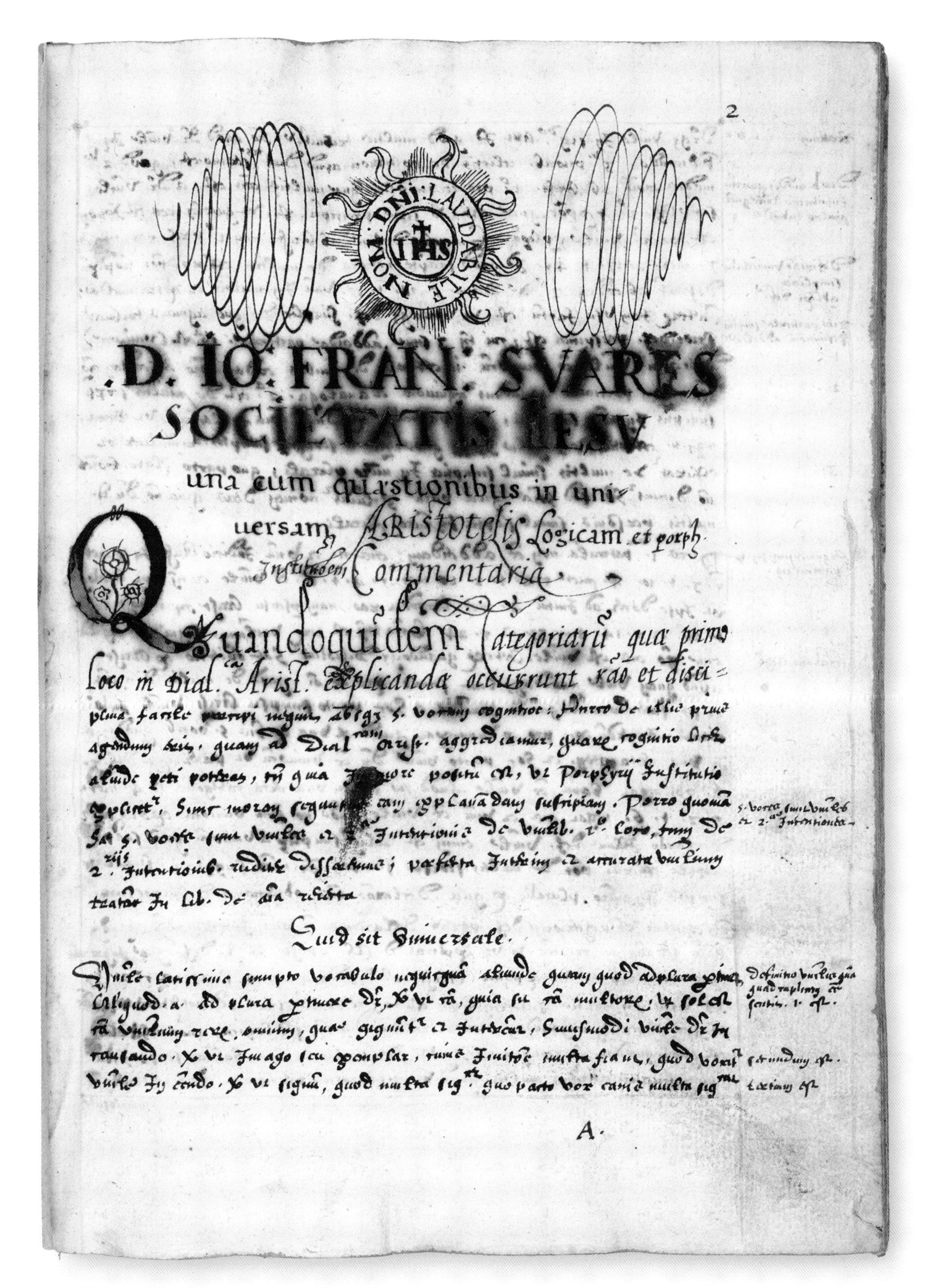

Cat. 8. First page

9

SCHOOL NOTES FROM SPAIN

Unidentified Spanish commentator

[Commentary on Aristotle's *Categories* and *Posterior Analytics*, and Porphyry's *Isagoge*]

Spain, 1606
Manuscript on paper

From the twelfth century, Aristotle's Organon—a collection of his six works on logic—was the foundation text for training in logic in the West. The Iberian Peninsula was home to some of the most important Aristotelian centers: Toledo, the royal court where the first translations from Arabic into Latin had been promoted, and Salamanca and Coimbra, where the universities were of consistent distinction in the editing and understanding of the entire Aristotelian corpus. Often in manuscripts and in printed editions Aristotle's works of logic would be accompanied by the *Isagoge*, or "Introduction," written in Greek by the philosopher Porphyry in the third century CE, and soon translated into Latin. It remained for centuries a standard manual of logic, providing a clear classification of concepts often graphically represented as the Tree of Porphyry.

This manuscript appears to be notes from a set of lectures in a school. It was written by a Spanish or Portuguese scholar, who dated it 1606. On the final page he transcribed a quote from Aristotle as a sort of motto: "It is absurd at the same time to know and to seek a method for knowing," an apt tag for an age with an increasingly scientific attitude to knowledge. Although the writer does not reveal his identity, the binding on the book suggests that he regarded his work as something worthy, an important achievement. He went to the trouble and expense of commissioning a rather costly vellum cover decorated in gilt. The motifs point perhaps to a connection with Jewish circles. It would have been a brave statement: Jews had been expelled from Spain in 1492, and *conversos* (Jews who had converted to Christianity), while allowed to remain, were mistrusted, discriminated against, routinely tried, and often convicted.

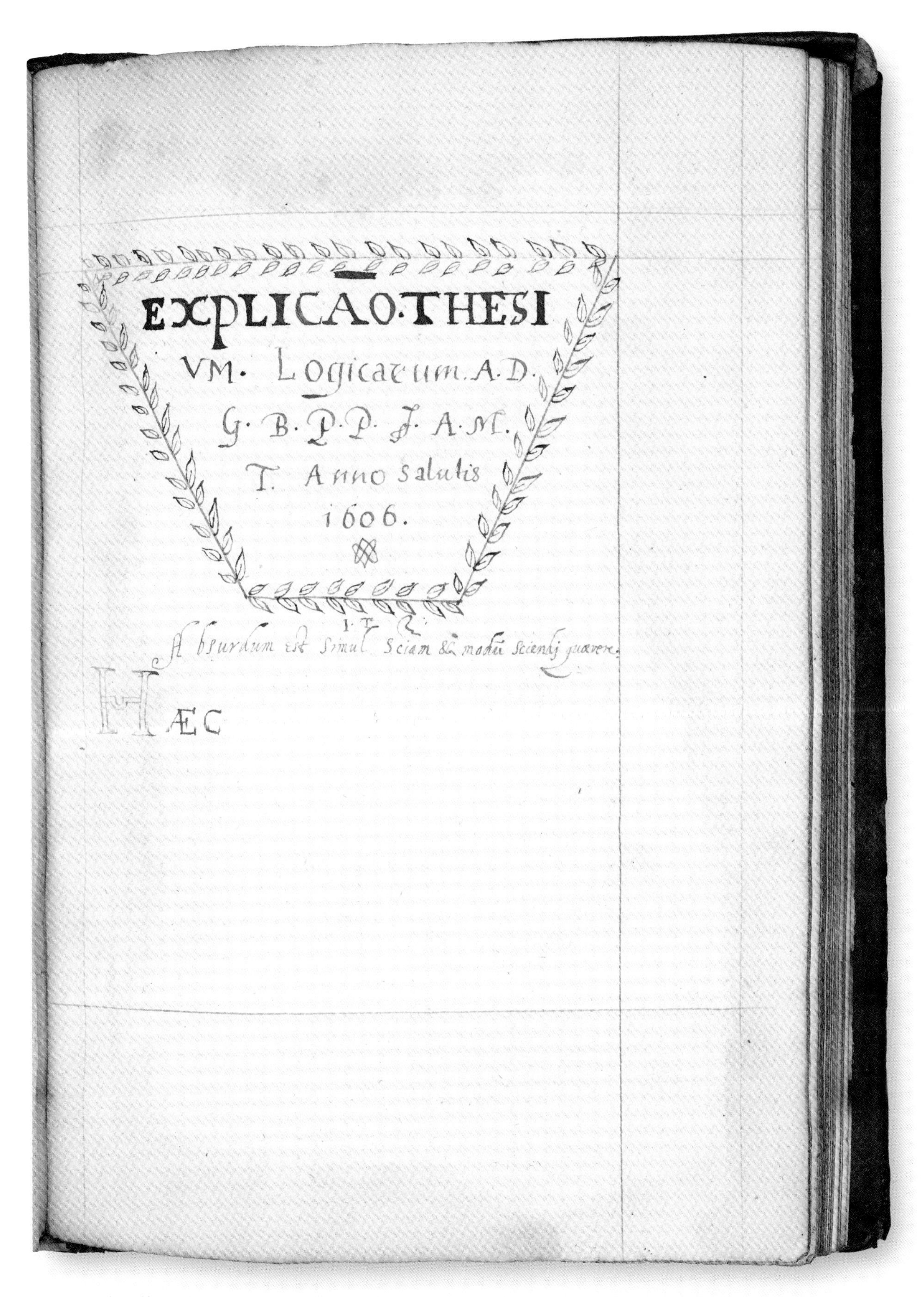

EXPLICAO·THESI
VM· Logicarum. A.D.
G·B·P·P· J·A·M·
T. Anno salutis
1606.
I·T·
Absurdum est simul sciam & modū sciendi quærere.
HÆC

Cat. 9. Explicit (from the Latin meaning "here ends..."), containing the title of the manuscript, the initials of the author, and the date

10
SPANISH ANNOTATORS

Pierre Tartaret (d. 1522)

Commentarii in libros … de anima … philosophiae naturalis [et] metaphysice … in sex libros ethicorum

Venice: Sessa and Petrus de Ravanis, 1520
Annotated by two Spanish or Portuguese scholars (c. 16th century)

Pierre Tartaret taught in Paris, and had been known since the 1490s as an authority on the works of Aristotle. His commentaries, spoofed by Rabelais, were used as university textbooks; they were collected and printed together as a body after 1500. This volume contains Tartaret's expositions of Aristotle's logical, ethical, metaphysical, and scientific works, together with his commentaries on the logic of the Spanish medieval physician (and later pope) Peter of Spain.

The book is thoroughly annotated by at least two contemporary hands, both likely to be Spanish or Portuguese. The first reader, responsible for the majority of the notes, shows a particular interest in Aristotelian logic. He is especially interested in philosophical vocabulary and offers alternative definitions. He also takes a keen interest in other commentaries, such as those by the medieval logicians Duns Scotus (1265/66–1308) and Albert of Saxony (c. 1320–1390). The second annotator, writing with a tighter hand, concentrates mostly on the nature of sensation and its relationship to knowledge. There were two main schools of Aristotelian thought on the Iberian Peninsula: the Calculatores, who, following the Oxford Calculators, developed refined theories of motion and speed, and the later Neo-Scholastics, such as Francisco Suárez and the Coimbra group, deeply committed to logic (cat. 8). These schools played leading roles in modernizing Aristotle, adapting him to the interests and concerns of the early modern period.

Naturalis inclinatio duobus modis accipitur actualiter & sic [illegible] realiter a suo subiecto [illegible] in aia & determinataz ad [illegible]

Prohemium II

A ¶ Questiones admodum subtiles ⁊ utiles / cum medulla totius materie artium / quatuor librorum sententiarum ⁊ quotlibeti doctoris subtilis Scoti in suis locis quotate / magistri Petri Tatareti parrhisiensis super libros logices Porphyrii ⁊ Aristotelis cum textus clarissima expositione / ac dubiorum seu difficultatum ordinatissima terminatione feliciter incipiunt.

Logica an sit scientia ab alijs distincta.

Circa initiũ totiꝰ logice mouet talis q̃stio: vtrum logica sit scia rõnalis vna ⁊ ab alijs distincta. ¶ Arguit pᵒ q non sit scia: qr ois scia acqrit p demõstratione: sz logica nõ acqrit p demõstratione: g ipa nõ est scia. ¶ Maior pz ex diffõne scie. ¶ Minor pbat: qr si logica acqreret p demõstratione aň logicã eēt alia logica, p quã illa demõstratio eēt nota: ⁊ sic eēt pcessus i infinitũ. ¶ Scdo sic: Modus sciẽdi nõ ẽ scia: sz logica ẽ modus sciẽdi: g logica nõ ẽ scia. Maior pz: qr i sfm acqrẽdi sciaz nõ est ipa scia. Et minor pbat p Aris. 2. metaphysice dicẽtẽ: q absurdũ est sil qrere sciaz ⁊ modũ sciẽdi. ¶ Tertio sic, pbãdo q ñ sit rõnal: qr qlz scia ẽ de ens reale existẽs i aia de pnto qlitatis: g nlla erit scia rõnalis, cũ reale ⁊ rõnale opponant.

B ¶ Quarto sic, pbãdo q nõ sit vna scia: qr illa scia nõ est vna in q̃ colligunt plures pclones demõstrabiles p diu'sa media: sz sic ẽ de logica: g nõ ẽ vna. Maior pz: qr scia ẽ hitꝰ pclonis: g mltiplicat pclonibꝰ mltiplicant ⁊ scie. ¶ Quinto sic, q ñ sit ab alijs distincta: qr illa scia nõ ẽ ab alijs distincta q̃ ẽ de aliq̃ obto cõi cũ alijs scietijs: sz logica est hmõi: igit. Minor pz: qr logica ẽ de syllo vel mõ sciẽdi, q̃ vtũt oẽs alie scie: g nõ ẽ ab alijs distincta. ¶ In oppᵐ argr p Phm tradẽtẽ sciaz logicalẽ vnã ⁊ ab alijs distinctã. ¶ In qõne ista ⁊ i oibꝰ alijs erũt tres articuli. In pᵒ dclabũt imi qõnis. i 2ᵒ mouebũt dubia i 3ᵒ ponet pclo rñsal ad q̃sitũ / ⁊ soluẽt rões aň oppᵐ.

Logica duplex est.

Primo sciẽdũ q pro declõne huius termini logica / supponit q logica est duplex. s. nãlis ⁊ artificialis. Logica nãlis nõ est aliõ q̃ nãlis iclinatio vel q̃dam habilitas nãlis nãliter inexistens aie vel itellectui, qua nãlr inclinat vel est potens ad assentiendũ magis vero q̃ falso: ex quo assensu pcedimus de noticia noti ad noticiam ignoti. ⁊ de ista ponunt alique ppões. ¶ Prima: ista logica nãlis est idem realr cũ aia vel itellectu: ⁊ hoc capiendo ipsaz pro iclinatione ipsius aie vel itell's: qr ois iclinatio est eadẽ realr cũ illo cuius est iclinatio. Et dr notãter, capiẽdo p iclinatione: qr si capet pro habilitate ipsius aie vel itell's, posset bñ distigui ab aia ⁊ itellectu: cũ habilitas ⁊ ihabilitas sint q̃li-

Logica nãlr qd sit.

Sc. li. 3. di. 17. ⁊ i. 4. di.

ates de scda specie qualitatis. ¶ Scda ppositio: logica nãlis nõ est scia: ideo p ipsaz nõ dicimur sciẽtes: sz p ipsaz possumus dici logici. ¶ Tertia ppõ: oẽs sumus iclinati nãlr ad assentiẽdũ vero, tñ aliq magis vel minus sunt dispositi: iõ aliq pmptius ⁊ facilius assentiũt vero q̃ alij. Sz logica artificial' ẽ duplex. s. doces ⁊ vtens. Doces est que docet diffinire diuidẽ arguere ⁊ ver a falso discernẽ: ⁊ ista acqrit p demõstratione, ⁊ p certas reglas artis logice. Sz logica vtẽs capit duobꝰ modis. Unoᵒ p noticia sciẽtifica logice acqsita p demõstratione q̃ vtimur in singulis sciẽtijs diuidẽdo, diffiniendo, arguẽdo, ⁊ ver a falso discernẽdo: ⁊ sic est vera scia non distincta a logica docẽte. Alioᵒ capit p habitu vel noticia acqsita ex freqnti argumẽtatiõe ⁊ exercitio arguẽdi: siue talis noticia fuerit vã siue falsa siue dubia vl certa, siue cũ formidie siue nõ: ⁊ sic logica vtens nõ est vã scia, sz tñ pmptitudo q̃dã ad tales actus exercẽdos gñata ex frequẽti argumẽtatione: qr fm Sco. ex oi actu volũtatis vel itell's põt gñari habitus vl pmptitudo: qr talis hitus gñat in manu. ¶ Adũte vlterius q logica artificial' seu doces põt capi duplr. Unoᵒ pro simplici hitu alicꝰ pclonis logicalis acqsito p demõstratione vel demõstrationes. Et sic quilz hitus vel noticia alicꝰ pclonis demõstrabilis in logica põt dici logica artificial', ⁊ sic põt describi: Logica artificialis est scia vnius pclonis logical' p demõstratione vel demõstrationes acqsita. Et ab ista nõ dr cõiter ⁊ vsualr hõ logicꝰ: qr ad h q aliqs dicat cõiter logicus, opz q hẽat oẽs noticias oium pclonũ logicalium vel saltẽ maiorẽ partes. ¶ Et siqs dicat: qlz noticia scientifica põt aliqs dici sciens: ergo p illum habitũ qs erit sciens, ⁊ non nisi scia logica: igit tal' erit logicus. Rñdet q aliqs p talẽ habitũ vnicꝰ pclonis pot dici logicus, sz nõ vt cõiter vtimur logico. Alioᵒ capit logica artificialis p collectiõe pluriũ noticiarũ adhesiuarũ logicaliũ: ⁊ h triplr. Unoᵒ large, p aggregato ex oibꝰ noticijs adhesiuis tã pncipioꝝ logicalium q̃ pclonũ ex eisdẽ pncipijs deductaꝝ siue scitaꝝ siue opinataꝝ, i. siue sciant cũ formidine siue non. Alioᵒ capit stricte pro aggregato ex oibꝰ noticijs oium pncipioꝝ logicaliũ ⁊ pclonũ scitaꝝ deductaꝝ ex illis pncipijs. Tertio mõ capit magis stricte. s. p aggregato ex noticijs adhesiuis conclusionũ logicaliũ demõstrataꝝ p pnᵃ logicalia. Et de ista ppe dat diffinitio logice, cuz dr q logica est scia docẽs diffinire diuidẽ, arguere ⁊ ver a falso discernere p rões: ⁊ talem distinctiõne debes pportionabilr capere in q̃lz scia. ¶ Aduerte finalr q capiendo logicã artificialẽ 3ᵃᵒ adhuc est duplex. s. vetus ⁊ noua. Vetus / q̃ determinat de partibus argumentationis ⁊ sylli tã ppings q̃ remotis: ⁊ pp hoc dr vetus ⁊ nõ pp antiqtatẽ: ⁊ ista ptinet sub se librũ pdicabilium, pntoꝝ pihermenias, librũ diuisionũ Boetij, librũ supponum ampliationũ, restrictionũ, appellationum, ⁊ distributionum. Alia est logica noua q̃ determinat de syllo, ⁊ de ptibus sbiectiuis eius: ⁊ ista ptinet sub se librũ prioꝝ, posterioꝝ, topicoꝝ, elenchoꝝ, ⁊ ẽt tractatum pñaꝝ, insolubiliuz, ⁊ obligationum.

A 2

L sti. 49. q. 10. in corpore qõnis ⁊ sol. arg.

Sc. li. 3. dis. 33.

Iste cõditiones cõueniũt hitui: vt declarat. Sco. li. 1. di. 17. q. 2.

ponitur in naturali potentia

Notuz hic sumitur pro propositione uera vel magis uera, ignotuz pro propositione falsa vel dubitabili

Conclusio logicalis dicitur propositio in qua predicatur proprius alicuius subiectuz de eo vel in qua ponitur aliquis terminus qui est de consideratione logices, sicut sunt termini importantes 2ᵃˢ intentiões

Cat. 10. Introduction with both marginal and interlinear annotations

11

READING A COMMENTARY

Giovanni Pietro Apollinare Offredi (15th century)

Expositio in primum Posteriorum Aristotelis

Venice: Bonetus de Locatellis for Ottavianus Scotus, 1493

The area in northern Italy called Lombardy was a vibrant hub of culture in the early Renaissance, home to the refined Milanese courts of the Visconti and the Sforza, who promoted contemporary artists and thinkers, and to the prestigious University of Pavia. As an esteemed philosopher, Giovanni Offredi was active both at court and in the academy. Little is known about Offredi; but he did leave this commentary on Aristotle's *Posterior Analytics,* a keystone of logic in which Aristotle teaches the reader how to arrive to correct conclusions (that is, conclusions proven correct by facts as they stand in the material world) by correctly linking valid premises. We know that Offredi's book was read and valued for over a century because Galileo used it himself when he set out to write his own commentary on the same work by Aristotle; it is therefore an important, if little-known, link in the chain of the transmission of Aristotle's text.

This work is also important because an unknown scholar added his notes in the margins, probably only a few years after the book had been printed. Much as Galileo would have done a century later, he engages with Offredi's interpretations, especially those passages that talk about conditions for producing true knowledge.

Posteriorum

detur tripliciter probare licz obscure. pmo sic. Scire est rē p cām cognoscere ⁊ quoniā illius est causa ⁊c. Et scire est rē p demonstrationem intelligere. ģ demonstratio ē ex pmissis veris imediatis ⁊c. tenet ꝯña. qꝛ dato opposito ꝯsequentis seqtur oppositum alteri⁹ pmisse. ⁊ antecedēs p vtraqꝫ pte ex precedentib⁹ patet. cum includat duas diffinitiones ipsius scire. Hanc rōnem tangit cum dicit. Si quidē igitur est scire vt posuimus. ❡ Scđo arguit sic. illud qđ est ex p prys pricipys eius quod demōstrat est ex primis veris immediatis ⁊c. sed demōstratio est hmōi. ergo ⁊c. patꝫ ꝯseqntia cum maiori. ⁊ minor nota est. qꝛ aliter demonstratio nō generaret sciam. ❡ Tertio arguit sic. ois sillogismus faciens scire est ex premissis veris. sed demonstratio est huiusmodi. ģ ⁊c. minor apparet ex precedēti diffinitione demonstrationis. maior vero tenet. qꝛ ille vidētur eē ꝯditiones sillogismi faciētis scire. ⁊ ideo dicit phs ꝙ sine predictis cōditionibus cōtingit fieri sillogismum absolute nō tñ erit demōstratio cū nō gñabit sciam. ❡ Nōndū est pmo ꝫ linco. cū i textu dicat demr̃atiua scia ē ex pmis. ⁊c. ꝙ pt dupl' littera itrodici. Uno mō p demr̃atiuā sciaꝫ dituz scientificuz itelligēdo. sic videnẗ sonare vba. ⁊ sic s'm illū sensū non ponit diffinitio noua demr̃atōis sed solū infert illa ꝯclo ex p dictis. ꝙ scia p demr̃atiōem acqsita est ex pmis veris ⁊c. ⁊ tūc ista ppositio ex vt ingt linco. denotat cām efficientē q̃ simul ē efficiēs ⁊ origo. sic. n. pater ē cā efficiēs ⁊ origo fily. ita notitia primoꝝ veroꝝ ⁊c. ē efficiens ⁊ origo notitie scientifice ꝯclonis. Scđo mō pōt itroduci. vt p illū terminū demr̃atiua scia demr̃atōeꝫ itelligam⁹. ⁊ sic ꝯñr noua ꝯcludit diffinitio demr̃ationis. ⁊ tūc illa ppō ex dic̃ sil' circūstātiaꝫ cause mālis ⁊ cause efficiētis. qđ p tāto dixit linco. qꝛ q̃daꝫ pticule i illa diffōne posite significant ea q̃ formal' demr̃atione igrediūt. ⁊ p ꝯñs sūt ptes ⁊ mā ipsi⁹. qre q̃tū ad illas pticulas ly ex dic̃ circūstātiā eē mālis. Quedā vo ea q̃ solū vtual' ipsā igrediunẗ. ⁊ iō quo ad illas ly ex denotat cāꝫ efficētē. Ex quo patꝫ ꝯtra quosdā modernos ꝙ h scđa diffinitio nō solū est sūpta a cā māli ꝫ a māli ⁊ efficiēte sil'. ❡ Notandū scđo ē ꝙ pma diffinitio demr̃atōis sūpta ē a cā finali cū dr̃ demr̃atio ē sill's faciēs scire. eo ꝙ scire ē finis demr̃atōis vt oēs dicūt. ⁊ qꝛ finis ē cā causaꝝ dād' alys causis necessitatē imponēs ⁊ prior ipis. iō diffinitio pma demr̃atōis. diffinitio ipsi⁹ formalis dici debet. ❡ Scđa vero diffinitio mālis. qre dū scđa p pmā pbatur illa pbatio est a priori. qꝛ diffinitio mālis eiusdē rei p formalem demonstratur.

VErum quidem igitur oportet esse quoniam quod non est nō est scire. vt ꝙ diametros sit simetros.

Simeter

Dyameter

❡ Ondit pticulas diffinitionis pdicte eē cōuenienter positas. ⁊ pmo facit hoc q̃ ad illam pticulā veris. scđo quo ad illā pticulaꝫ primis ⁊ imediatis. ibi. Ex pmis aūt. Tertio quo ad illā pticulam prioribus ⁊ notioribus causis ⁊c. Primo igit pbatur ista conclusio. demr̃atio ē ex veris. arguēdo sic. ꝯclo demr̃atōis nō scit̃ nisi p premissas. cū igit p demōstrationeꝫ acqrat scia ꝯclonis. seqtur ꝙ ēt premisse sciuntur. sed nō ē pos̃e illud qđ nō ē. i. falsū sciri. vt vbi gr̃a. ꝙ diameter q̃drati sit ꝯmēsurabilis sue coste. q̃re ⁊c. ꝯña ⁊ añs vidētur manifesta. ❡ Circa lr̃am dubitant cōiter exponētes ꝙ Arist. vr̃ facere ꝯñaꝫ illaꝫ. illō quod demr̃atur ē veꝝ. ģ pmisse demr̃ationis sūt vere. ꝫ illa ꝯña nō vr̃ tenere. vt patet i logica. lꝫ eni sequatur. illa ꝯña ē bona. ⁊ añs est verum cū alys circūstātys q̃ habēt alibi videri.. ģ ꝯñs est verum. nō tamē vr̃ seq ecōtra. qm̄ ex falsis pōt veꝝ iferri. Ad hoc pmo dr̃ ꝙ predicta ꝯña nō exp̄sse colligitur i textu. sed textus vr̃ inuere illū modū exponendi q positus fuit. ❡ Dicitur scđo. ꝙ siue illā ꝯñaꝫ fecisset phus siue non ipsa cōcedēda est gr̃a mãe non de forma. qꝛ i demr̃atōeꝫ pp qd de qua loqmur non solū premisse sunt causa ꝯclonē iferendi. imo ēt dn̄t causam eēntialeꝫ significati ꝯclusiōis in essendo. falsum aūt qđ nō est i reꝝ natura ⁊ qđ ē tanq̃ puatio nō potest esse eēntialis causa rei necessarie q̃le est significatiuū ipsi⁹ ꝯclonis. qre ⁊c. ❡ Nōndū ē pp exēplū adductum ꝙ quadratoꝝ quoddam ē habēs quatuor angulos eq̃les ⁊ rectos. ⁊ tale proprie dr̃ quadratum. quod ēt orthogonium eqlaterum dici solet. dr̃ autem orthogoniū ab orthos quod est rectuꝫ. ⁊ gonos angul⁹ quasi quadratū rectorum anguloruꝫ. ❡ Aliō ē q̃dratū habens āgulos equales sed latera ineq̃lia. sicut duo maiora ⁊ duo miora. qđ orthogonium ieqlaterū vocari pt. Aliō est habēs angulos iequales et latera iequalia quod ēt dici potest iorthogonium eqlaterum. Quartum vero est habens latera iequalia ⁊ angulos iequales. Hec tamē tria iprop̄e quadrata dñr. sed vnumquodq; ipsoruꝫ debꝫ q̃drāgulum appellari. ❡ Ad ppositū igr̃ dico ꝙ de q̃drato proprie sumpto verificatur exemplū phi non autem de alys. Primus declarat̃. nam ille linee cōmensurabiles dñr qb⁹ assignari potest aliqua mēsura cōmunis que vtraq; illaruꝫ reddat aliquotiens sumpta. ⁊ ecōtrario illa dñr icōmensurabilia. Talia vero sunt dyameter ⁊ costa quadrati. nulla eni mensura aliquotiēs sumpta dyametrū adequate mensurat que aliquotiens sūpta eius costam adequate mensuret. neq; ecōtra. nam si sic esset tunc dyameter ad costā eēt aliqua pportio rōnalis qualis est numeri ad numeruꝫ. sed nulla talis inter ipsa repitur vt demr̃atur quarto euclidis.

Quadratum ortogonium eqlateꝝ

Quadratū orthogonū ieqlaterū.

Quadratū eqlaterū nō orthogoniū.

Quadratum neq; orthogonium neque equilaterum.

Cat. 11. Quadrilaterals illustrating definitions of geometric concepts

12

A LATE COMMENTARY ON THE SOUL

Jacobus Salomonius (17th–18th century)

Commentaria et quaestiones in tres libros Aristotelis De anima ad mentem Doctoris Angelici D. Thomae Aquinatis

[Padua, 1655]
Manuscript on paper

This manuscript must have been an important object to its compiler and owner: the title page is profusely decorated in ink wash and watercolor with a floral and zoomorphic frame, and the text is carefully penned. It contains an otherwise unknown commentary on the philosophy of mind, based on Aristotle's *On the Soul*, written in 1655 by a professor at the University of Padua, where Galileo, Cesare Cremonini (cat. 21), and the main protagonists of the new science in northern Italy had staged their momentous controversies only a few years earlier. This manuscript contains notes on Salomonius's course taken by a student named Ercole Avogadro. The commentary is a witness to the post-Galileo stage of the interpretation of Aristotle's thought on the nature of knowledge, linking his mapping of science with a new preoccupation with the nature of the human mind. It was this issue that, at this time, claimed almost the entirety of, to quote just the most outstanding example, Descartes's dedication. Thus, much attention is devoted to the relationship between perception and consciousness: the unity of the intellect is seen as the necessary condition and the guarantor of the exactness of the knowledge derived from perceptions.

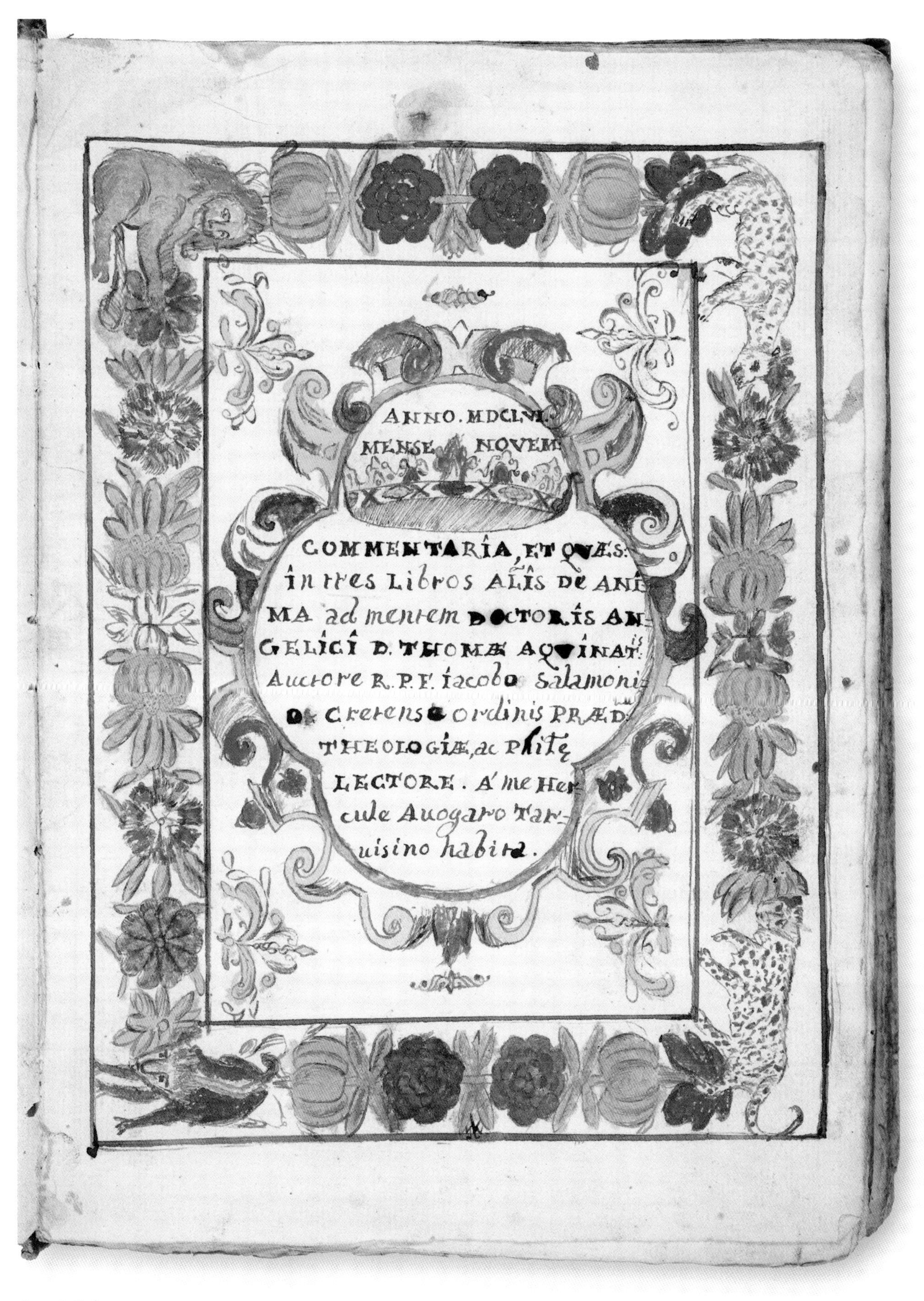

ANNO. MDCLVI.
MENSE NOVEM.

COMMENTARIA, ET QVAES:
in tres Libros Aris de ANI-
MA ad mentem DOCTORIS AN-
GELICI D. THOMAE AQVINAT:is
Auctore R.P.F. Iacobo Salamoni-
o Cretense ordinis PRAED:
THEOLOGIAE, ac Phitę
LECTORE. A me Her-
cule Auogaro Tar-
uisino habita.

Cat. 12. Title page

Meoeu[illegible] sobij novarius i 4'' ☾u° [illegible]

exemus ☾ i 7 i 6 [illegible] ℔. xepus ions 47 [illegible]

more.

Typus Defectus Solaris Partialis

THE NATURAL WORLD

"We think we know a thing when we know its primary causes and its primary principles, all the way up to its elements."

PHYSICS, BOOK I, CHAPTER 1

ARISTOTLE PRODUCED A HUGE BODY of original research in physics, astronomy, chemistry, and zoology. He held that all things tend towards a certain end or goal, or *telos*. Teleology is the study of the ends of things. We understand nature by looking at how forces either set objects in motion or send them towards rest. To the four natural elements—fire, earth, air, and water—which change and mix and experience "generation and destruction," Aristotle added aether—that is, the celestial world—which is unchanging and moves eternally in crystal-like spheres, hence the largely predicable "movement" of the stars. Aristotle's experiential knowledge led him to hold the Earth to be spherical, and this influenced science for many centuries.

Translations of and commentaries on Aristotle's works on physics, astronomy, and zoology filled monastic as well as academic libraries, and were used as springboards for studies on light, for mathematical applications, for measurements, and for predictions. In the early modern era, when evidence coming from the new experimental science threatened to dethrone Aristotle's model of an Earth-centered universe, the very Aristotelian demand for coherence, systemic order, and reliability soon forced theory to catch up with observation and experiment.

John Manners (1609–1695)
Disputationes in octo libros Physicorum
Perugia, Italy, 1647
(detail of cat. 28)

13

MEDIEVAL TESTIMONY

Aristotle

De physica

France, 13th century
Manuscript on vellum
Annotated by an unidentified scholar (13th century)

This single leaf is from a decorated medieval manuscript on vellum written in Latin around the third quarter of the thirteenth century. It contains a translation of part of book VI of the *Physics* and must have been transcribed from a high-quality, accurate university copy. It is an eloquent witness to the rediscovery of Aristotle in Spain and Italy, where the first translations from Arabic into Latin were made in the twelfth century, and to the spread of Aristotelian texts throughout Europe in the thirteenth century.

The material from which this leaf was made is animal skin: parchment or vellum (the two terms are often interchangeable in this context). The preparation of vellum was a slow and complex process, involving weeks of soaking the skin in fresh water, then lime and water, followed by drying, scrubbing, and rinsing to obtain a clean and hair-free surface. (We can often still easily detect the side where the hair was—the "grain side.") At this point, the skin was stretched on a frame and fixed with adjustable pegs. As it dried, the skin would shrink and tear at the edges, and any small flaws in the skin would be pulled out into holes. When this happened, writers and copyists would simply work around the holes, so they did not affect the completeness of the text. More scraping followed, producing a thinner, finer, and thus more desirable, parchment. An expensive process makes for an expensive product, but parchment is exceptionally durable, lasting for more than a millennium without decaying. What better support for a text that was meant to accompany mankind throughout history?

This is a wonderful example of a medieval scholarly manuscript, as visually effective and economical as the most accomplished of today's web pages. The writer has made extensive use of abbreviations and textual compression to maximize the expensive parchment, but also adopted colored paragraph marks and running titles to ease the reader's orientation; a near-contemporary annotator then inserted contrasting interlinear additions with alternative readings and well-defined, immediately recognizable large blocks of commentary in the margins. The effect is evident. On a page that,

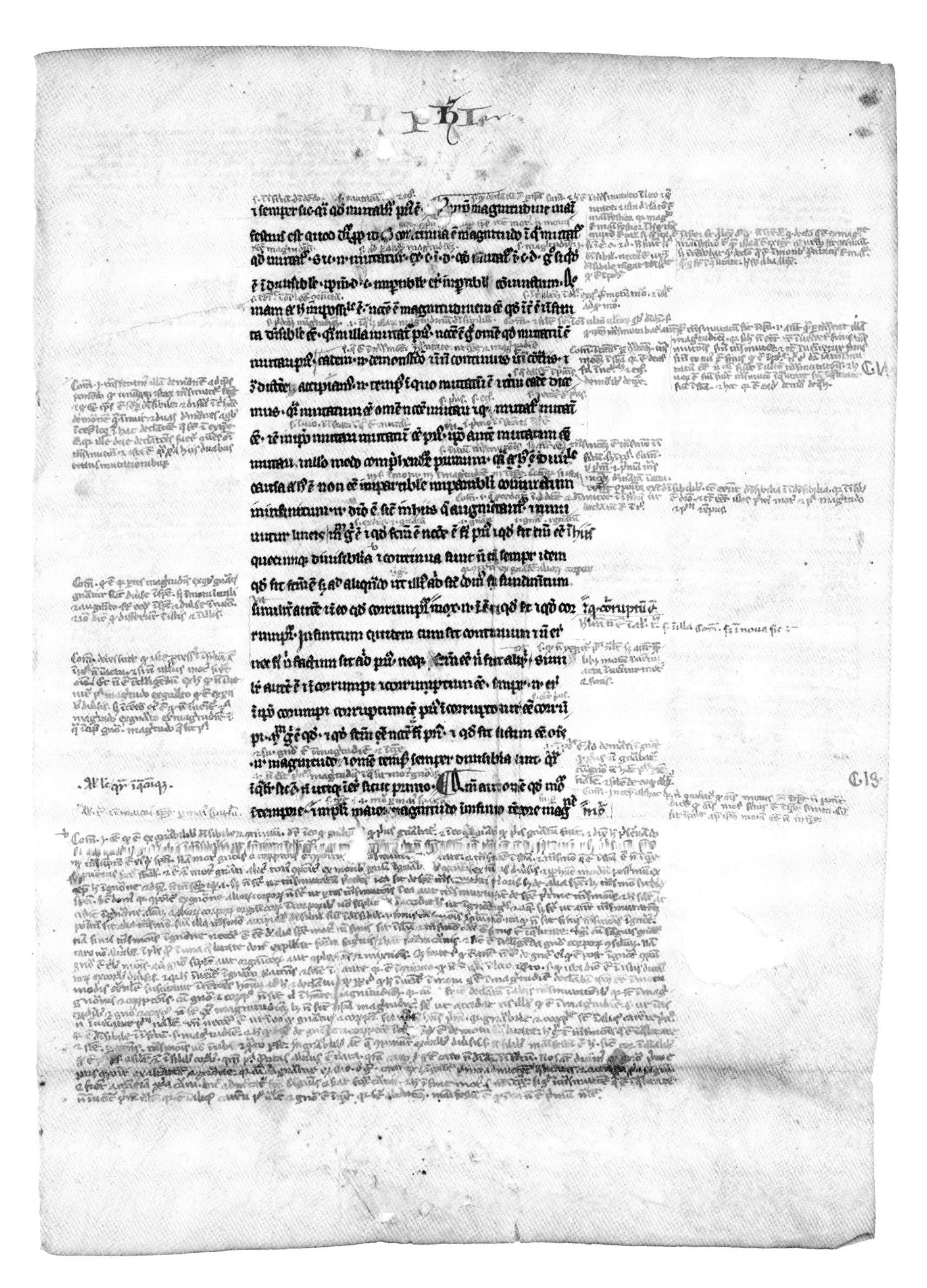

Cat. 13, recto. The Annotations occupy five different fields according to their content: between the text lines, in three columns in the margins, and in the lower margin

without careful planning, would have been impossibly crowded, the sizes and positioning of the various scripts allow the eye to scan and recognize the main text, central and larger, then choose to “click the link” to the alternative readings, or the interpretation, or the commentary.

14

THE UNIVERSITY OF COLOGNE

Gerhardt de Harderwijk (d. 1503)

[Commentary on Aristotle's *Physics*]

Germany, c. 1480
Manuscript on paper

At the time this commentary on the *Physics* was written, the University of Cologne in Germany was one of the greatest centers of Aristotelian thought in Christian Europe. Albert the Great (d. 1280) had been the first to tackle the entire Aristotelian corpus and give it an interpretation that was in harmony with Christian teachings. It was in Cologne that the other great systematic thinker of the Middle Ages, Thomas Aquinas (1225–1323), heard Albert's lessons and developed a friendship with him. Albert showed that, since Aristotle's philosophical study of the physical world was based on experience, it was a separate domain from theology, which was based on divine revelation. He therefore believed in the pursuit of both domains—philosophy and the natural sciences on one hand, and theology on the other, each with its own method. The University of Cologne thus became a hub of philosophy, and the natural sciences were pursued with few theological shackles.

Gerhardt (or Johannes) de Harderwijk was one of Albert's successors at Cologne and continued his predecessor's exegesis. His work enjoyed great popularity in the schools. It gave, amongst other things, a complex account of the cognitive process, in which the brain is divided into several functional areas, and knowledge results from the interaction between external objects, sensory perception, and brain faculties, including memory. The version in this early manuscript contains significant variants absent from the printed edition, and this makes it of special importance for scholarly research.

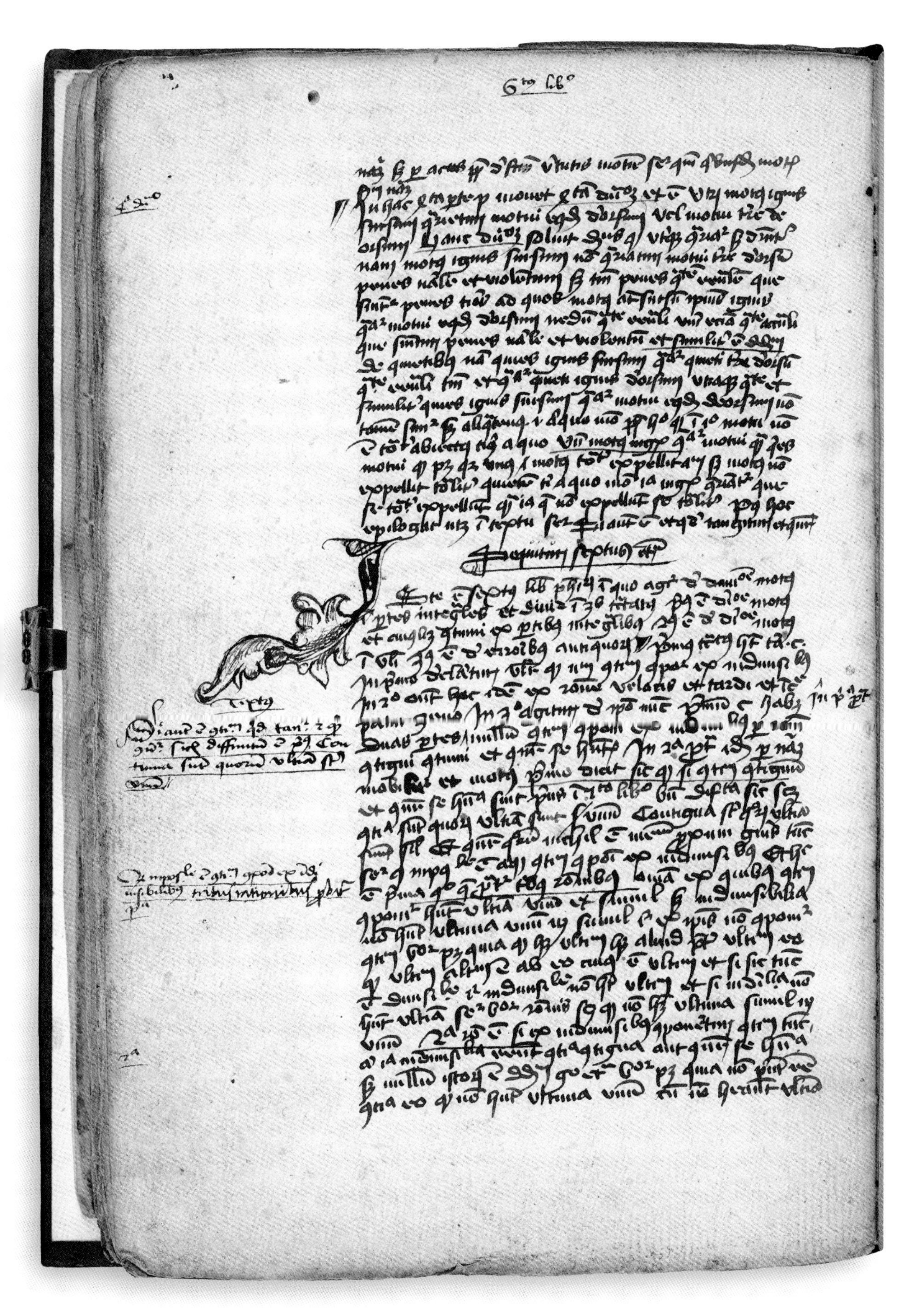

Cat. 14. Flourish indicating a section break

15
AN IMPORTANT DOMINICAN COMMENTATOR

Johannes Versor (d. c. 1485)

Quaestiones super VIII libros Physicorum Aristotelis (cum textu). [With:] Quaestiones iuxta textum De anima Aristotelis (cum textu)

Cologne: Heinrich Quentell, 13 March 1497 and 5 September 1496
Annotated by two readers (c. late 15th century)

The typical academic question-and-answer form frames these two commentaries. The author, Johannes Versor (the Latinized name of Jean Letourneur), was a prominent French Dominican who was appointed rector of the University of Paris in 1458. We know very little about him, but it is becoming clear that his role in making Aristotle's thought relevant in the early Renaissance was significant. Starting in 1485, his interpretations of Aristotle enjoyed a broad circulation in Europe, including in Spain. By the late 1490s they had been almost completely translated into Hebrew by the Spanish philosopher Eli ben Joseph Chabillo (Elijah ben Joseph Habillo). They are seen by some as the source for much sixteenth-century Hebrew philosophy, such as that of Baruch Ibn Ya'ish.

Annotated by two early readers, these two commentaries were clearly bound together at an early stage and near the place of printing, very probably by their first owner. Putting together a study of the *Physics* with a study of *On the Soul* (*De Anima*) would have been typical of a follower of the Aristotelianism proposed by Averroes (cat. 3). For Averroes, the study of the soul, or psyche, was part of the realm of physics, since the psyche is an essential part of the union of form and matter found in the physical world.

Prohemiū Physicoꝝ

Physicis errores primus docet anteriores.
Scribit principia. que docet esse tria.

Circa initium primi libri phisicoꝝ. Queritur Utrū de rebus naturalibus et physicis sit scientia. ¶ Arguitur ꝙ nō tribus rōnibus Eracliti q licet diceret de omnibꝫ nihil posse sciri. hoc tñ de naturalibus seu physicis maxime asseruit Prima rō est Entia physica seu naturalia nō vere sciuntur nisi sciantur ſm suū esse ⁊ ſm ꝙ sunt. sed ipa hoc modo capta infinitas hñt drās vel saltē eis nō repugnat habere ipas ſʒ intellectus noster refugit infinitū nec ab eo cōprehendi pt. ergo entia physica seu naturalia ſm suum esse et ſm id q̄ d sunt ab intellectu nostro cognosci nō pñt ⁊ ꝑ ɔñs neqꝫ sciri. Scda rō. si entia physica scirentur hoc esset per demonstratōem cuius definitio esset medium. sed in naturalibꝫ definitio est ipius ꝑticularis que nō potest esse mediū demonstratōnis. Minorem probat ipe eraclitus sic. qꝛ si naturalia in cōmuni accepta habeant definitōnem erit equiuoca eo ꝙ ratio animalis ſm vnumquodqꝫ animal est altera et altera. ⁊ talis definitio est inutilis ad demonstrandū ergo de physicis nō est scientia Tercia rō. entia physica seu naturalia sunt mutabilia. nam si consideretur forma naturalis in ordine ad materiam semp est mutabilis. ergo cū scia nō sit nisi de immobilibꝫ et necessariis sequitur ꝙ de entibꝫ physicis ⁊ naturalibꝫ nō potest esse scia. Et ista ratio eraclito et ericlitanis videbatur esse ita violenta. cum videbāt omne quod transmutatur ꝑtim esse in termino a quo ⁊ ꝑtim ī termino ad quem concedebant. contraria et cōtradictoria eidem inesse et de eodem verificari. Si em album moueatur dum est inter terminos motus ꝑtim est album et ꝑtim nigrum. et sic contraria de eodem verificantur vt dicebant. Et quia nigrum est non album. Et album et non album contradicunt. ideo similiter dicebant contradictoria de eodem verificari. Sed isti decepti fuerunt non percipientes differentiam inter subiectum et accidēs. et etiam inter habere contraria ſm actum imperfectuꝫ et ſm actum completum et perfectum. Pro responsione

¶ Sciendum est primo. Qꝫ cum scientia sit habitus intellectus in eo existens oportet omne illud de quo est scientia esse intelligibile. Est aūt aliquod intelligibile per hoc ꝙ aliqualiter est abstractum a materia ergo ſm ꝙ aliqua diuersimode abstracta sunt a materia ꝑtinent ad diuersas scientias. ꝯ cum omnis scientia habeatur per demonstratōem cuius medium est definitio oportet ſm diuersitates definitionum scientias diuersificari. ¶ Pro quo Sciendum est scdo Qꝫ res reales de q̄bus est philosophia realis differenter se habent ad materiam et ſm hoc earum est triplex definitio et pertinent ad tres partes phie realis.

a ij

Cat. 15. Introduction

16
GREEKS IN EXILE

Aristotle and Theodorus Gaza (translator, d. 1475)

De animalibus

Venice: Johannes and Gregorius de Gregorii, 1492
Annotated by RB (late 15th or early 16th century)

With the fall of the Byzantine Empire to the Ottoman Turks in the fifteenth century, Greek scholars took refuge in Italy and there launched a second great wave of Aristotelian scholarship in Europe. The works of the best Greek scholars were sponsored by princes and popes. Theodorus Gaza, who had emigrated to Pavia, then Naples, and finally to Rome, was celebrated by his contemporaries as an exceptionally fine translator. His rendition of Aristotle's *De Animalibus* (a combined volume of *History of Animals*, *Parts of Animals*, and *Generation of Animals*), of which this is an early edition, was famous throughout Europe and enjoyed a virtual monopoly for a long time. Legend has it that when Pope Sixtus IV, who had commissioned this work, rewarded Theodorus with several gold coins for the sublime beauty of the translation, as well as for the gold binding in which it was presented, the scholar disdainfully threw the coins into the Tiber.

In his imposing work on the animal kingdom, Aristotle catalogues the features of over five hundred animal species with an immense collection of biological data: anatomical, physiological, and behavioral. The late-fifteenth- or early-sixteenth-century annotator of this copy lavishes attention especially on the passages relating to reproduction in mammals and reptiles. He also leaves a tantalizing, as yet unresolved, clue as to his identity in the monogram *RB* written inside the woodcut blank coat of arms at the foot of the beautifully decorated title border.

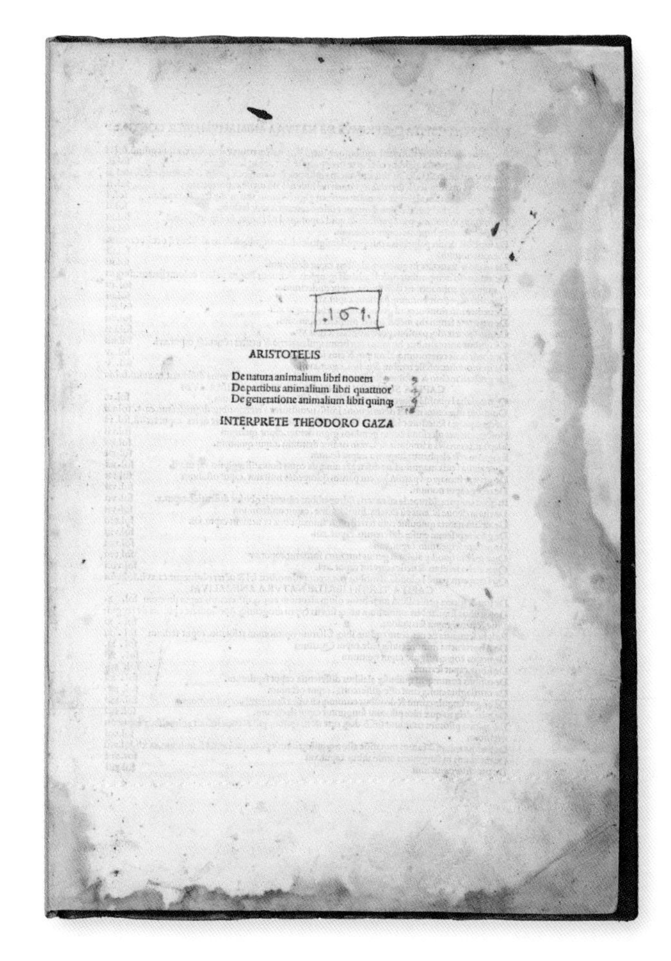
.161.

ARISTOTELIS

De natura animalium libri nouem
De partibus animalium libri quattuor
De generatione animalium libri quinqꝫ

INTERPRETE THEODORO GAZA

Cat. 16. Contents arranged on an early form of a title page, which only became a standard feature of bound books in the 16th century

THEODORI:GRAECI:THESSALONICENSIS:PRAEFATIO:IN LIBROS:DE ANIMA LIBVS:ARISTOTELIS:PHILOSOPHI:AD XYSTVM:QuARTVM PONTIFICEM MA XIMVM.

L Y curgum lacedemoniũ qui leges ciuibus suis constituit. Reprehendunt nõ nulli.
Pontifex summe Xyste quarte:ꝙ ita tulerit leges ut belli potius q̃ pacis rationẽ ha
buisse uideretur. Numam uero pompilium regem Romanum laudant maiorem ĩ
modum:ꝙ pacis adeo studiosus fuerit:ut nulla causa moueri ad bellum pateretur;
quorum sñiam & si alias probo:ut debeo(nihil enĩ pace cõmodius:nihil sanctius)
Tamen cum uita hominum ita ferat:ut bella uitari interdum nequeant. Sic censeo
prefiniendum cousulendumq; ut & bellũ interdum sit suscipiendum:si res urget:
& pax seruanda sit semper:si fieri potest:nec belli ratio unquã probanda sit:nisi ut demum rebus
compositis quieto tranquilloq; animo uiuamus. Non enĩ ad pugnam & homicidia: nõ ad discor
dias & bella nati sumus:sed ad cõcordiam & humanitatem :Itaq; principis institutũ atque officiũ
id esse reor ut pacẽ summa opera petat:seruet:& colat. Quod cum Romanos põtifices fere omnes
fecisse quo ad potuerint:intelligam:laudo illorum animum. Quod neq; ab instituto nature bone
recesserint:& preceptum auctoris diuini seruarint:quod sepissime pacem conciliat: & cõmendat.
Sed usum non nullorum ausim reprehendere. Pace enim qua uti debuerant ad litterarum & artiũ
bonarum studia:& uirtutum officia:illi quidẽ ad uoluptates parum honestas abusi sunt quod cũ
omni hominum ordini sit turpe:tum pontificis persone turpissimum est. fuerunt tamen & qui re
cte pace uterentur:& pontificatum magna cum laude gerent:quibus te similem uideo plãe succes
sisse .prestas enim doctrina & moribus:quo fit ut nomen tuum immortalitati mandandum .cen
seas studio potius litterarum quæ nunquam peunt:q̃ uel edificiorum que breui tempore delentur
uel thesauri gemmarum:& auri:cuius heredes inimici plerũq; succeduntq; uoluptates :que ani
mi lumen extinguũt respuas:unamque litterarum uoluptatem:que animum illustrare potest:expe
tas :atq; in ea requiescas:itaq; omnes pontificem talem habuisse lætamur:& quisq; siue ciuis siue p
egrinus studet pro uiribus:rem afferre: qua & uirtutum tuarum laudem:& animi tui uoluptatem
augeat:quem enim ob eius merita non solum ueneramur & obseruamus : sed etiam perinde ac
patrem amamus :ei perlibenter opera studiorum nostrorum dedicanda putamus:alii igitur alia.
Ego tibi libros aristotelis philosophi quos de natura animalium scripserat:Conuertere in latinũ
sermonem uolui:existimans rẽ tibi gratam me facturum. Si hi libri latinis litteris mera industria
mandarentur:& cultius atq; integrius:quam actum ab aliis adhuc est exponerentur:uidebam per
multa errasse interpretes tum imperitia linguæ :tum aristotelicæ disciplinæ inscitia. Itaq; officiũ
esse putaui:ut si quid facultatis in me esset :qua prestare hæc possem meliora: rem aggrederer:& p
uiribus dignius quicquam efficere conarer. Laboraui equidem hac in re uehementer quoniã
nihil adiumenti ab iis qui eadem interpretari uoluere:capere poteram. Aut enim græce illi dixe
runt quæ latine audire homines latini desiderãt:aut rerum aliarum nomina aliis improprie tribue
runt:aut noua ipsi inepte finxerunt. Sententiam uero auctoris passim adeo deprauarunt : ut argu
mento quidem interpretationis eorum nemo tam indoctus sit:qui non rectius iudicare de rebus
naturæ:quam aristoteles posse uideatur. Locutionis etiam genus ne tale quidem quod legi posset:
non modo non aristotelis eloquentia dignum :adhibere potuerunt. Quod si quid præ sua facilli
ma cognitione peruerti ab omni ueritate non potuit:id etiam propter incultum horridum & ine
ptum sermonem interpretis uix ĩtelligi potest. Afferrem hoc loco scripta illorum interpretum :er
roresq; singillatim enumerando reprehenderem:nisi longior essem in re non dubia :presertim a
pud te princeps doctissime:qui ante doctor pro tuo singulari iudicio semper damnasti illoꝝ inter
pretationem:melioremq; desiderasti:sic nunc factus princeps accuratius agis:ut hæc aliquando ex
planentur:& una cum sui auctoris sententia:lucem latinæ linguæ recipiant. Vtinam tantũ faculta
tis in me esset priceps sapientissime.ut hoc meum ĩterpretandi genus :esse optimũ illud quæris cõ
firmare possem. Sed quoniam hoc plus est:q̃ ut uel homo modestus de se polliceri debeat:uel meæ
uires attingere possint:nihil ego tale de me :illud uero dicere licet:hanc esse oĩbus quæ adhuc fac
tæ sunt Horum libroꝝ ĩterpretationem :& philosophis ad eloquendum cõmodiorem:& eloquẽ
tibus ad disputandum utiliorẽ & utrisq; ad aristotelis sẽsũ percipiẽdũ aptiorẽ. Nec acta equidem
egi:sed a male cõuersis primus ego latinis hoĩbus apui.q̃ his libris aristoteles scripserat:sed illuc re
deo. Me plurimũ elaborasse in his libris ĩterpretãdis fateor. Cũ nihil a primis ĩterpretibus illis iuua
ri possẽ.sed oĩa ex codicibus ueteꝝ auctoꝝ petere necesse haberẽ:lectõne lõga:notatiõeq; uaria pli
niũ. Corneliũ:columellã:uarronẽ:catonẽ.M. Tulliũ:apuleiũ:gelliũ:senecã:cõpluresq; alios liguæ
latiæ auctores euoluere diligẽtius oportuit.quoꝝ libros uel semel accurate legisse laboriosũ ẽ:ne
dũ ad usũ ĩterp̃tãdi accõmodasse:nõ eni ita usũ uerborũ accipit ĩterpres:ut suo arbitratu rẽ dispo
nat mandetq; elocutioni:sed totidẽ fere uerbis sẽsus aliẽos accõmodate exprimere cogitur : qđ

Cat. 16. Preface

17

A MAJOR EARLY OXFORD COMMENTARY

Robert Grosseteste (1175–1253)

Summa Linconiensis super octo libris Physicorum. Expositio Sancti Thome super libros Physicorum Aristotelis

Venice: Petrus de Quarengiis for Alexander Calcedonius, 22 April 1500

Robert Grosseteste, first chancellor of Oxford University and bishop of Lincoln, is one of the most important scholars in medieval England. He wrote what is believed to be the earliest commentary on the *Posterior Analytics* in the West, which served as the foundation for his philosophical and scientific work. His commentary on the *Physics*, constructed out of notes written in his hand in the margin of his copy of the text, is believed to have been finalized about 1235. This was Grosseteste's main work in natural philosophy, and its interpretation of Aristotle proved fundamental to the development of medieval science. This copy is from the first printing.

Grosseteste was particularly curious about the nature of light, which he believed to be the key to understanding the various fields of knowledge. However, in the present commentary he puts forward the idea of "unequal infinities." Grosseteste had complete faith in measuring as a method for attaining knowledge. He was fond of pointing to the Bible passage that says that God "has disposed all things in number, weight and measure" (Wisdom 11:21), and notably refuted Aristotle's arguments in the *Physics*, book VIII, for a world without a beginning or end. Perhaps the most important legacy he left to his immediate successors in science was a strong emphasis on the importance of mathematics.

Diui Roberti Linconiensis super octo libris physicorũ breuis & utilis summa feliciter incipit.

In primo libro physicorum cuius subiectũ est corpus mobile simpliciter non contractum. Determinat phũs de principijs corporis mobilis ipsum constituentibus: & sunt principia nãe siue corporis mobilis tria. s. materia forma & priuatio. Materia est illud q se habet ad formam sicut es ad statuam & lignum ad lectum: & silr sicut informa se hẽt ad formam anteq̃ ipsaʒ recipiat. vel materia est illud ex quo fit res & manet in re. sp materia ẽ in quã vnũquodqʒ compositum nãle resoluitur. Proprietates mãe sunt tres. prima materia non est scibilis nisi per comparatio-

Cat. 17. Beginning of the commentary

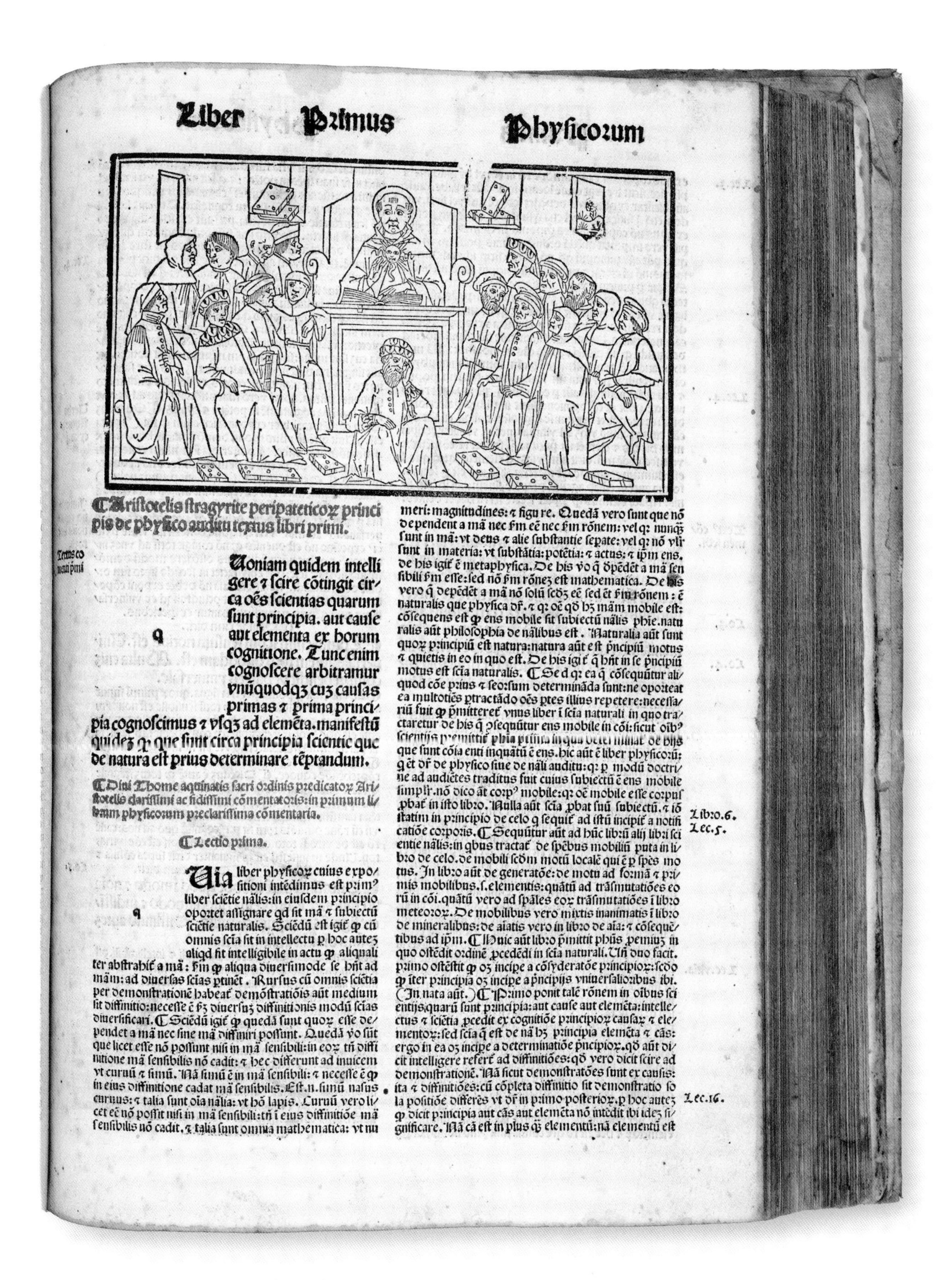

Liber Primus Physicorum

¶ Aristotelis stagyrite peripateticorum principis de physico auditu textus libri primi.

Textus co[mmenti] primi

Quoniam quidem intelligere et scire cõtingit circa oẽs scientias quarum sunt principia. aut cause aut elementa ex horum cognitione. Tunc enim cognoscere arbitramur vnũquodqȝ cuȝ causas primas et prima principia cognoscimus et vsqȝ ad elemẽta. manifestũ quideȝ q[uod] que sunt circa principia scientie que de natura est prius determinare tẽptandum.

¶ Diui Thome aquinatis sacri ordinis predicatorum Aristotelis clarissimi ac fidissimi cõmentatoris: in primum librum physicorum preclarissima cõmentaria.

¶ Lectio prima.

Quia liber physicorum cuius expositioni intẽdimus est primus liber sciẽtie nãlis: in eiusdem principio oportet assignare qd sit mã et subiectũ sciẽtie naturalis. Sciẽdũ est igit q[uod] cũ omnis scia sit in intellectu p[er] hoc auteȝ aliqd fit intelligibile in actu q[uod] aliqualiter abstrahit a mã: s[ecundu]m q[uod] aliqua diuersimode se hñt ad mãm: ad diuersas scias ptinẽt. Rursus cũ omnis sciẽtia per demonstrationẽ habeat demõstratiõis aũt medium sit diffinitio: necesse ẽ s[ecundu]ȝ diuersuȝ diffinitionis modũ scias diuersificari. ¶ Sciẽdũ igit q[uod] quedã sunt quorum esse dependet a mã nec sine mã diffiniri possunt. Quedã vo sũt que licet esse nõ possunt nisi in mã sensibili: in eorum tñ diffinitione mã sensibilis nõ cadit: et hec differunt ad inuicem vt curuũ et simũ. Nã simũ ẽ in mã sensibili: et necesse ẽ q[uod] in eius diffinitione cadat mã sensibilis. Est.n. simũ nasus curuus: et talia sunt oĩa nãlia: vt hõ lapis. Curuũ vero licet eẽ nõ possit nisi in mã sensibili: tñ ĩ eius diffinitiõe mã sensibilis nõ cadit. et talia sunt omnia mathematica: vt numeri: magnitudines: et figure. Quedã vero sunt que nõ dependent a mã nec s[ecundu]m eẽ nec s[ecundu]m rõnem: vel q[uia] nunq̃ȝ sunt in mã: vt deus et alie substantie sepate: vel q[uia] nõ vlr sunt in materia: vt substãtia: potẽtia: et actus: et ip[su]m ens. de his igit ẽ metaphysica. De his vo q̃ dpẽdẽt a mã sensibili s[ecundu]m esse: sed nõ s[ecundu]m rõneȝ est mathematica. De his vero q̃ depẽdẽt a mã nõ solũ scdȝ eẽ sed ẽt s[ecundu]m rõnem: ẽ naturalis que physica dr. et q[uia] oẽ qd hȝ mãm mobile est: cõsequens est q[uod] ens mobile sit subiectũ nãlis phie. naturalis aũt philosophia de nãlibus est. Naturalia aũt sunt quorum principiũ est natura: natura aũt est pncipiũ motus et quietis in eo in quo est. De his igit q̃ hñt in se pncipiũ motus est scia naturalis. ¶ Sed q[uia] ea q̃ cõsequũtur aliquod cõe prius et seorsum determinãda sunt: ne oporteat ea multotiẽs ptractãdo oẽs ptes illius repetere: necessariũ fuit q[uod] pmitteret vnus liber ĩ scia naturali in quo tractaretur de his q̃ ꝯsequũtur ens mobile in cõi: sicut oib[us] scientijs premittit phia prima in qua determinat de his que sunt cõia enti inquãtũ ẽ ens. hic aũt ẽ liber physicorũ: q ẽt dr de physico siue de nãli auditu: q[uia] p[er] modũ doctrine ad audiẽtes traditus fuit cuius subiectũ ẽ ens mobile simplr. nõ dico ãt corp[us] mobile: q[uia] oẽ mobile esse corpus pbat in isto libro. Nulla aũt scia pbat suũ subiectũ. et iõ statim in principio de celo q sequit ad istũ incipit a notificatiõe corporis. ¶ Sequũtur aũt ad hũc librũ alij libri sciẽtie nãlis: in qbus tractat de spẽbus mobiliũ puta in libro de celo. de mobili scdm motũ localẽ qui ẽ p[ri]ma spẽs motus. In libro aũt de generatõe: de motu ad formã et primis mobilibus. s. elementis: quãtũ ad trãsmutatiões eorũ in cõi. quãtũ vero ad spãles eorum trãsmutatiões ĩ libro meteororum. De mobilibus vero mixtis inanimatis ĩ libro de mineralibus: de aĩatis vero in libro de aĩa: et cõsequẽtibus ad ip[su]m. ¶ Huic aũt libro pmittit phũs pemiuȝ in quo ostẽdit ordinẽ pcedẽdi in scia naturali. Vñ duo facit. primo ostẽstit q[uod] oȝ incipe a cõsyderatõe principiorum: scdo q[uod] iter principia oȝ incipe a pncipijs vniuersalioribus ibi. (In nata aũt.) ¶ Primo ponit talẽ rõnem in oĩbus scientijs quarũ sunt principia: aut cause aut elemẽta: intellectus et sciẽtia pcedit ex cognitiõe principiorum causarum et elementorum: sed scia q̃ est de nã hȝ principia elemẽta et cãs: ergo in ea oȝ incipe a determinatiõe pncipiorum. qd aũt dicit intelligere refert ad diffinitiões: qd vero dicit scire ad demonstrationẽ. Nã sicut demonstratões sunt ex causis: ita et diffinitiões: cũ cõpleta diffinitio sit demonstratio sola positiõe differẽs vt dr in primo posteriorum. p[er] hoc auteȝ q[uod] dicit principia aut cãs aut elemẽta nõ intẽdit ibi ideȝ significare. Nã cã est in plus q̃ȝ elementũ: nã elementũ est

Libro.6. Lec.5.

Lec.16.

Cat. 17. Aquinas as magister with his students

18

A HUMANIST-SCIENTIST AT WORK

Simplicius of Cilicia (c. 490–c. 560 CE)

Commentarii in octo Aristotelis Physicae auscultionis libros

Venice: House of Aldus and Andrea Torresani, October 1526
Annotated by an unidentified scholar (15th century)

Along with his great rival Philoponus and Alexander of Aphrodisias (cat. 2), Simplicius of Cilicia ranks as one of the most profoundly influential commentators on Aristotle. He highlighted the blurred lines between Aristotelianism and Platonism. This book is the first appearance in print of his commentary on the *Physics*. It is the only text of its time to make reference to the thought of the Presocratic natural philosophers, who flourished in Greece in the sixth and fifth centuries BCE; their writings have been lost and would not be known at all had it not been for the citations in Simplicius.

The major issues in Simplicius's commentary are the nature of space, time, matter, and motion; moreover, he posited the existence of the void, which had been denied by Aristotle. These were subjects taken up repeatedly in early modern science, and they gave Simplicius's commentary a relevance that lasted through the seventeenth century. An anonymous annotator, probably an Italian humanist, exhibits a perfect understanding of the original Greek and pays particular attention to passages dealing with the key issue of kinesis—that is, "change," or "movement." His notes (written in an ink that has oxidized in places, resulting in a rather attractive sparkly finish) reveal a humanist-scientist at work, someone who, like Galileo and his teacher Francesco Buonamici (cat. 5), pored over an ancient commentary with the mind of a modern man well versed in the new experimental science. This volume was recently owned by Countess Antonia Ponti Suardi (1860–1938), an Italian noblewoman and philanthropist who founded a library in the town of Bergamo to promote the liberal and classical education of women.

ΣΙΜΠΛΙΚΙΟΥ ΥΠΟ-
ΜΝΗΜΑΤΑ ΕΙΣ ΤΑ ΟΚΤΩ ΑΡΙΣΤΟΤΕΛΟΥΣ ΦΥΣΙ-
ΚΗΣ ΑΚΡΟΑΣΕΩΣ ΒΙΒΛΙΑ ΜΕΤΑ ΤΟΥ ΥΠΟ
ΚΕΙΜΕΝΟΥ ΤΟΥ ΑΡΙΣΤΟΤΕΛΟΥΣ.

SIMPLICII COMMEN
TARII IN OCTO ARISTOTELIS PHYSICAE
AVSCVLTATIONIS LIBROS CVM IPSO
ARISTOTELIS TEXTV.

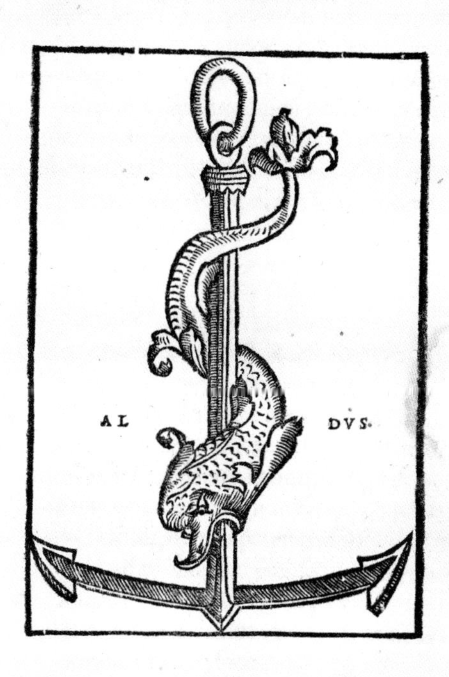

Ne quis alius impune, aut Venetiis, aut uſquam locorum hos Simplicii Cõmentarios imprimat, &
Clementis VII. Pont. Max. & Senatus Veneti decreto cau-
tum eſt.

Cat. 18. Title page

19

AN ENCYCLOPEDIA

Johannes Thomas Freig (1543–1583)

Quaestiones physicae

Basel: Sebastianus Henricpetrus, 1579
Annotated by Jacob Aigen of Marckdorf (16th century)

Before the sciences became specialized, universities valued books that claimed to encompass the sum of all knowledge—*pansophia*. This book by the German philosopher Johannes Thomas Freig, professor at the University of Freiburg, is a veritable encyclopedia. It covers a vast range of subjects, from astronomy, geography, hydrography, meteorology, metallurgy, pyrotechnics, optics, music, zoology, and botany, to dendrology, anthropology, and psychology. It relies on Aristotle, but also draws on Pliny, Livy, Agricola, Julius Caesar, Pindar, Suidas, Cicero, Martial, Galen, Plato, Virgil, Archimedes, and Plutarch. Moreover, it organizes its subject matter in a way that had been promoted by Petrus Ramus (1515–1572), Freig's anti-Aristotelian teacher, whose biography Freig had written. Freig's open homage to Ramus, an outspoken Protestant convert, lead to his expulsion from the university as a Protestant. Banned from teaching, Freig retired to Basel and worked as a proofreader for Sebastian Henricpetri (Sebastianus Henricpetrus), the printer of the present work. He died of the plague in 1583. This copy belonged to one Jacob Aigen from Marckdorf, who used it as a reference book. His many annotations afford unique insight into the ways in which early modern scholars could use Aristotle along with a wide repertoire of sources in their work.

Cat. 19. Binding

20

ALL IN THE FAMILY

Aristotle and Simone Porzio (translator, 1496–1554)

Aristotelis vel Theophrasti de coloribus libellus

Paris: Vascosan, 1549
Annotated by Guillaume Chrestien (1500–1558)

Tradition in its literal sense means “handing down.” Yet, as texts are handed down they can also be transformed, subtly adjusted to accommodate the needs and interests of the times. The present book is an fine example of how multiple hands and voices participate in a dialogue of textual transmission.

The treatise *On Colors*—variously attributed to Aristotle or another Greek philosopher, Theophrastus or Strato—sets forth a theory of color based on the assumption that all colors are derived from mixing black with white. This theory influenced all subsequent theories until Newton’s experiment with refraction in the seventeenth century.

The text has been translated from the original Greek into Latin by the Italian philosopher and medical doctor Simone Porzio (1496–1554). Porzio is traditionally associated with Pietro Pomponazzi (1462–1525), in spite of their different interpretations of Aristotle, because Porzio also denied the immortality of the soul. In this work, Porzio suggests that the real author of the treatise is Theophrastus and not Aristotle, a position largely accepted today.

The Aristotle-Porzio text has in turn been interpreted by Guillaume Chrestien (1500–1558), physician to the French kings Francis I and Henry II. He fills the margins of the book with his own thoughts, adding a wealth of scientific and medical information. From the inscriptions on the title page, we learn that Chrestien eventually passed this book on to his son Florent, who was tutor and then librarian to King Henry IV of France, and Florent in turn passed it to his son Claude, a bibliophile and collector. At least four thinkers, influential to varying degrees, left their successive marks on these pages, showing us how the transmission of tradition can generate new knowledge over time.

Cat. 20. Preface with annotations

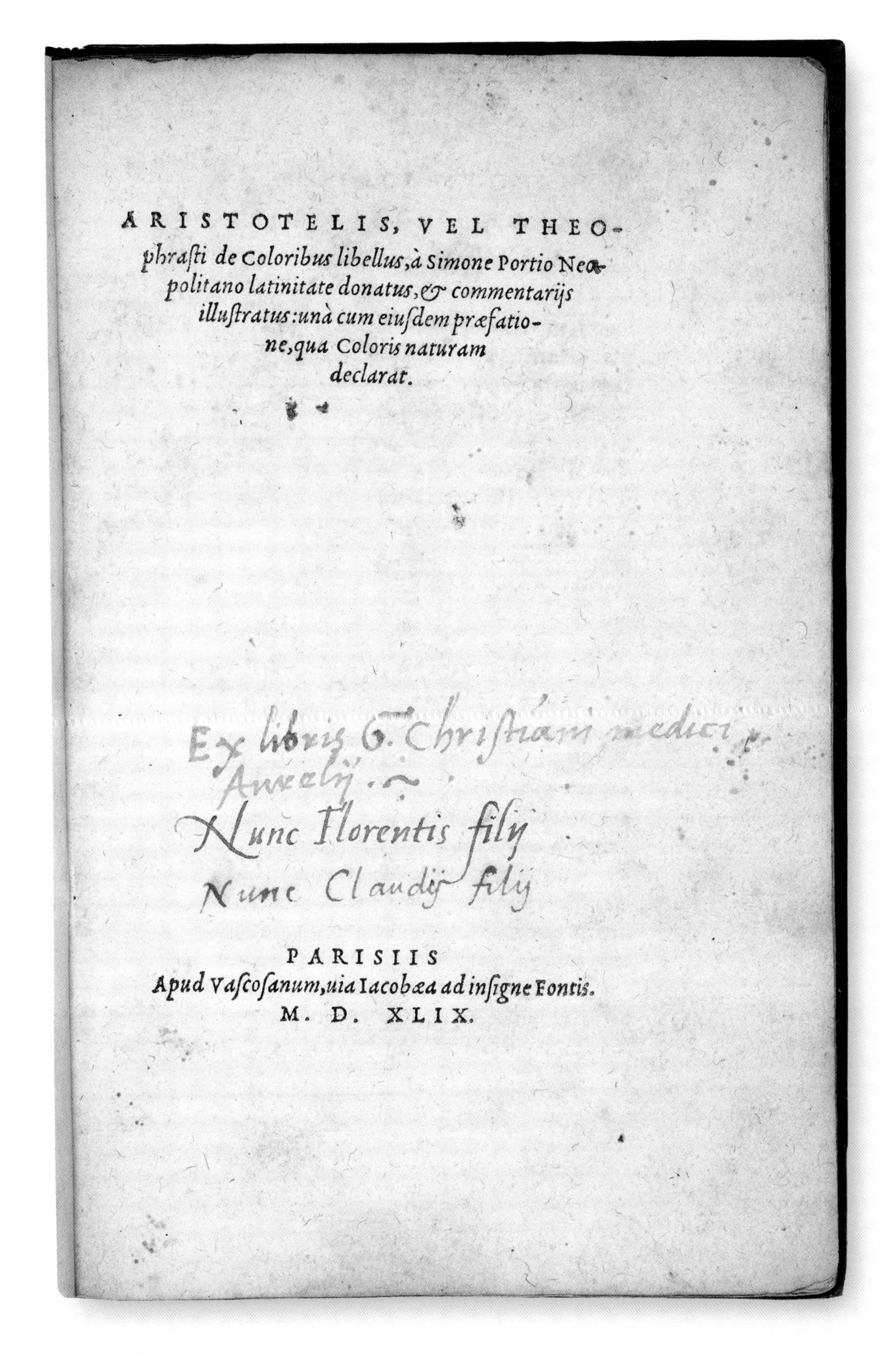

ARISTOTELIS, VEL THEO-
phrasti de Coloribus libellus, à Simone Portio Neapolitano latinitate donatus, & commentarijs illustratus: unà cum eiusdem præfatione, qua Coloris naturam declarat.

Ex libris G. Christiani medici Aurelij.
Nunc Florentis filij
Nunc Claudij filij

PARISIIS
Apud Vascosanum, uia Iacobæa ad insigne Fontis.
M. D. XLIX.

Cat. 20. Title page with inscriptions showing the succession of ownership

21

A CRITIC OF GALILEO

Cesare Cremonini (1550–1631)

Lectiones … super quartum meteorologicum

[Possibly Padua, c. 1598]
Manuscript on paper

Cesare Cremonini was a famed professor of natural philosophy at the University of Ferrara and the University of Padua. This manuscript contains two of his unpublished lectures. Cremonini counted among his pupils William Harvey, the English physician who gave the first account of the circulation of blood in the human body. Cremonini also had a relationship with Galileo, albeit a complex one: the two "natural philosophers" were at the same time personal friends and passionate scientific adversaries, each firmly standing on the opposite side of a deep divide. Cremonini was one of the two philosophers who refused to look through the telescope when Galileo announced its creation, and he resisted the potentially disruptive discoveries of the experimental new science. The knotty, rough-surfaced Moon that Galileo had seen through his telescope was incompatible with Aristotle's teaching, which Cremonini defended.

Cremonini's most important objection to Galileo's findings was his assertion that new discoveries and new data needed to be framed within a new theory that could hold them together coherently. Aristotle's all-encompassing system might be wrong, as Galileo's observations indicated, but, unless Galileo could provide an overarching theory of his own capable of accounting for the new data, his evidence would remain merely inconvenient. Cremonini knew that Galileo was likely to run into trouble with Church authorities, who protected an Earth-centered, man-centered view of the universe. He warned Galileo that moving to Tuscany would expose him to danger. Galileo went anyway.

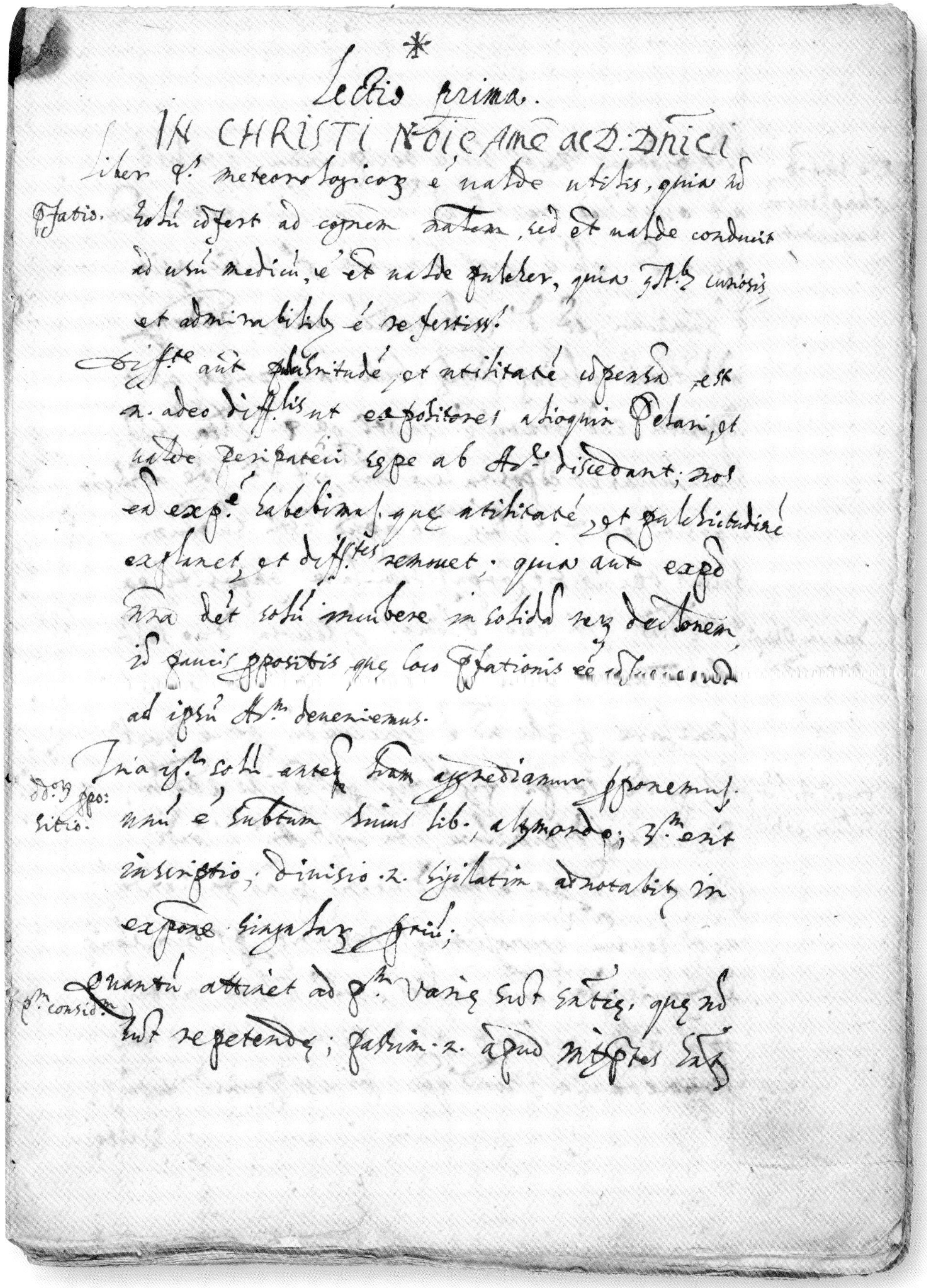

*

Lectio prima.

IN CHRISTI NOIE AME AC D: DNICI

Præfatio. Liber 4.us meteorologicorum ē valde utilis, quia nō solū confert ad cognitionem naturam, sed et valde conducit ad usū medicū, et ē valde pulcher, quia plurimis curiosis et admirabilibus ē refertissimus.

Difficultatem autem pulchritudinē et utilitatē conspersa est n. adeo difficilis ut expositores [illegible] [illegible], et valde Peripatetici saepe ab Ar.le discedant; nos ad expositionē habebimus, quae utilitatē et pulchritudinē explicet, et difficultates removet; quia autem expositio nostra debet solū incumbere in solida rerum declaratione, id paucis propositis, quae loco Præfationis esse [illegible], ad ipsū Ar.m deveniemus.

docendi propositio. Tria igitur solū anteque hunc aggrediamur proponemus. unū ē subiectum huius lib., alterum modo, 3.m erit inscriptio, divisio, et singulatim cum notabilibus in expositione singularum partium;

primum consid. Quantū attinet ad primū variae sunt sententiae, quae nō sunt repetendae; partim, et apud [illegible] [illegible]

Cat. 21. First page

22

JESUIT LECTURES ON THE PHYSICS

Giuseppe Agostini (1575–1643)

In octo libros Physicorum

Rome, 1605
Manuscript on paper

The author of this commentary on the *Physics* was the Jesuit Giuseppe Agostini, who taught at the Roman College (Collegio Romano) from 1603 to 1609. Professors there were central to the formation of Galileo's new science, and they proved to be steadfast supporters of the scientist and his discoveries. This manuscript, based on Agostini's lectures, was produced during his time at the college.

The physical features of the manuscript are important to note. The elaborate binding and the dedication to the Roman patrician Francesco Bufali mark its significance as an artifact fit for presentation to a patron of consequence. A provenance note dated 1636 testifies to its circulation during the dangerous years of Galileo's trial, when the possession of a text containing statements suspiciously in sympathy with Galileo, or challenging Aristotelian theories, would have created trouble.

Cat. 22. Book IV, chapter 1, "What Place Is"

Cat. 22. Book II, chapter 4, "Of Chance and Its Effects"

Cat. 22. Binding with the gilt armorial device of the noble Roman Bufali family

23

FRENCH COMMENTARIES

Anonymous

Commentarius in Physicam … in libros De caelo et elementis … in libros Meteorologicos … in libros De anima

France, early 17th century
Manuscript on paper

This manuscript is an extraordinarily thorough commentary on four works by Aristotle: the *Physics, On the Heavens, Meteorology*, and *On the Soul*. It was compiled by an accomplished French scholar, apparently before the publication and dissemination of Descartes's works, since the text contains no references to the French philosopher's influential and controversial ideas. From the extraordinary range of references in the manuscript, we can infer that the author had access to the best conceptual arsenal available to scholars of his age. He complements Aristotle's thought with ancient commentaries and independent works on the same subject, including references to Anaxagoras, Archimedes, Democritus, Heraclitus, Parmenides, Pythagoras, and Plato. He quotes the Church fathers (Ambrose and Augustine), the Islamic philosophers (Avicenna and Averroes), and the triad of the main scholastic philosophers (Aquinas, William of Ockham, and Duns Scotus). He collates these with the thought of more recent Aristotelians, such as Julius Caesar Scaliger (1484–1558) and Girolamo Cardano (1501–1576), and in discussions of theological matters he cites the major Protestant reformers Calvin, Luther, and Zwingli.

Cat. 23.
First page

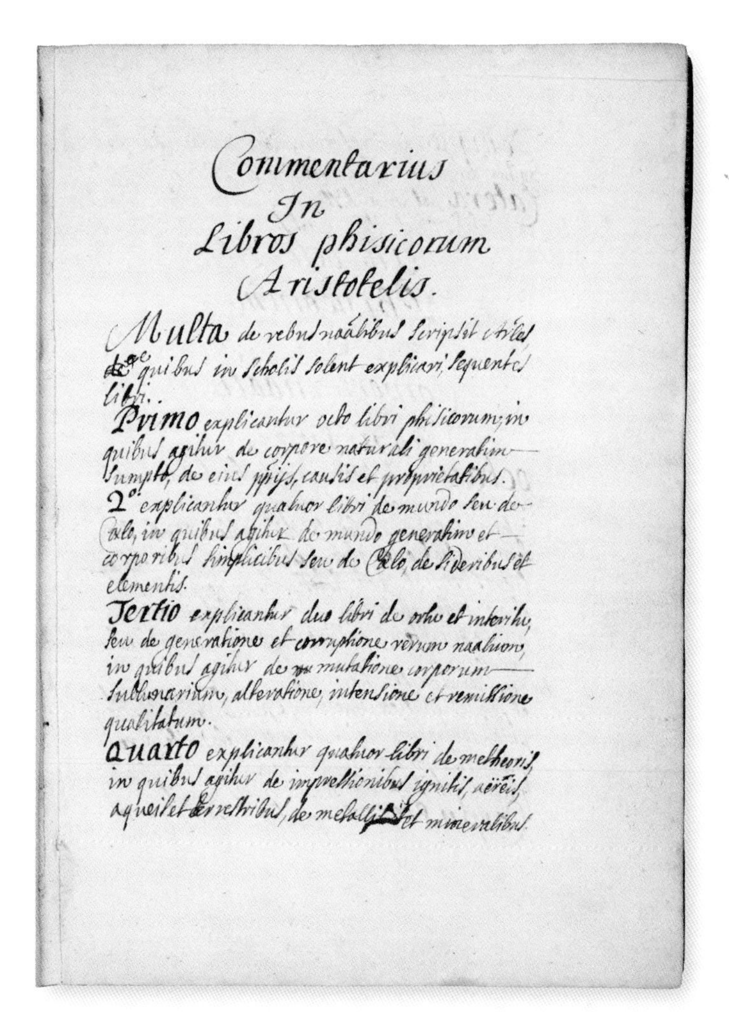
Commentarius
In
Libros phisicorum
Aristotelis.

Multa de rebus naturalibus scripsit Aristoteles de quibus in scholis solent explicari sequentes libri.

Primo explicantur octo libri phisicorum, in quibus agitur de corpore naturali generatim sumpto, de eius principiis, causis et proprietatibus.

2° explicantur quatuor libri de mundo seu de caelo, in quibus agitur de mundo generatim et corporibus simplicibus seu de caelo, de sideribus et elementis.

Tertio explicantur duo libri de ortu et interitu, seu de generatione et corruptione rerum naturalium, in quibus agitur de mutatione corporum sublunarium, alteratione, intensione et remissione qualitatum.

Quarto explicantur quatuor libri de meteoris in quibus agitur de impressionibus ignitis, aereis, aqueis et terrestribus, de metallis et mineralibus.

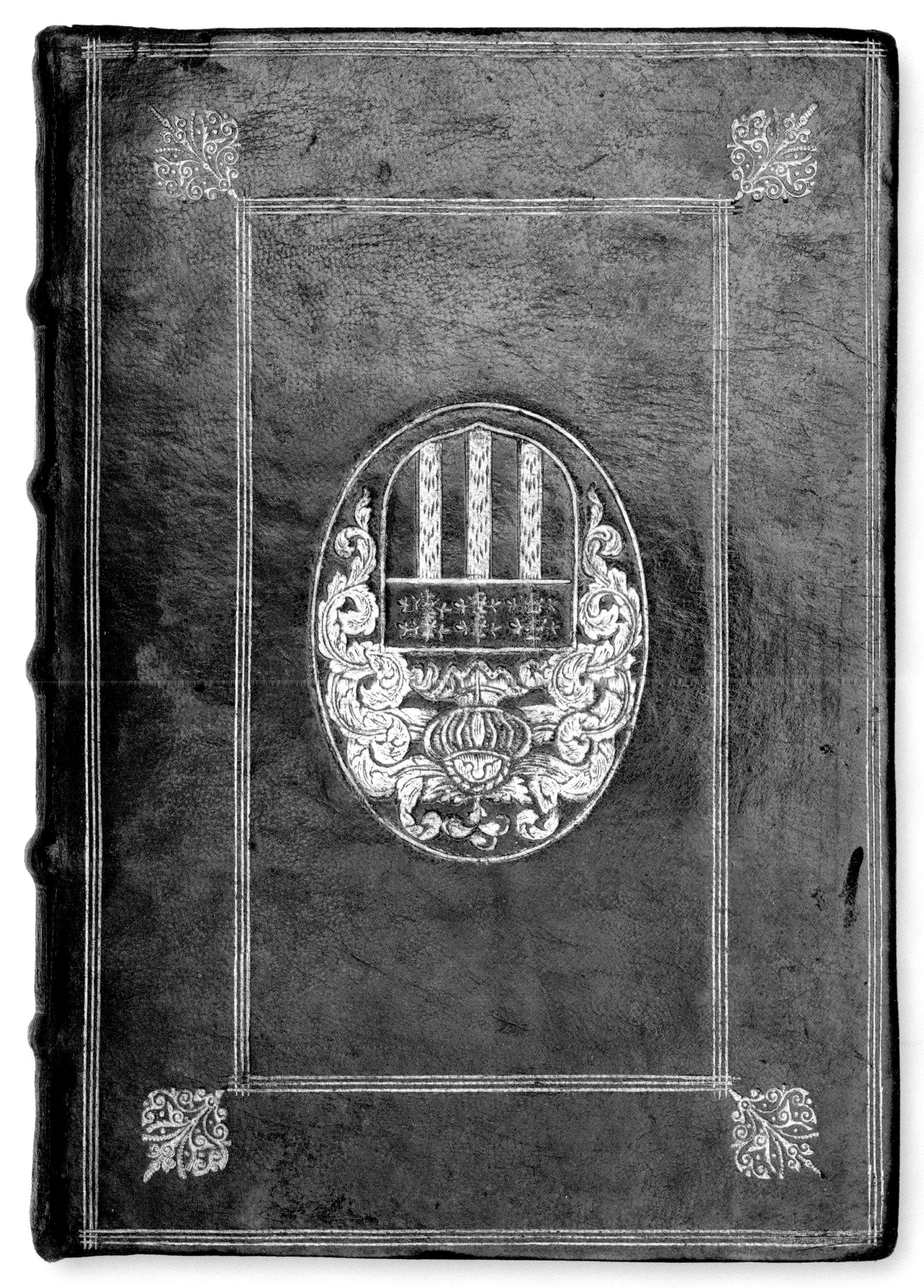

Cat. 23. Binding with gilt armorial device

24

STRADDLING THE OLD AND THE NEW

Alessandro Gottifredi (1595–1652)

In octo libros Physicorum … commentaria

Rome, 1629
Manuscript on paper

In the late sixteenth and early seventeenth centuries, the Jesuit Roman College (Collegio Romano) was culturally open and dynamic (cat. 8). After the publication of Galileo's *Sidereus Nuncius* (1610), the new science heralded by Galileo and his fellow thinkers produced results that, by their sheer incompatibility with the traditional religious outlook, could have driven the college, so close to the papal court, into an intransigent conservatism. After all, the Jesuits reported directly to the pope. How could Jesuit priests enter into dialogue with scientists who asserted that the Earth and mankind were not at the center of the universe; that the stars were not arranged forever in fixed concentric spheres around us; that the universe was much larger than previously assumed; that the Earth was but a small part of an enormous celestial space?

In spite of these challenges, this manuscript reveals a Jesuit school in touch with the latest developments in science and philosophy. It is a series of lectures on the *Physics* given by one of Francisco Suárez's successors at the college, Alessandro Gottifredi (1595–1652). Gottifredi enjoyed regular exchanges with a newly formed, independent group of philosophers and scientists who were not afraid to test the boundaries of orthodoxy in the name of experimental science: the Academy of the Lynx. Galileo, Federico Cesi (1585–1630), and Francesco Stelluti (1577–1652), among others, created the Accademia dei Lincei, naming it after the animal that, through its famed sharp powers of vision, symbolized the observational keenness required by the new experimental science. Gottifredi's lectures were intended for students from elite families who were introduced to the latest science through the safe gateway of Aristotle. The transmission of tradition here becomes at the same time an introduction to the new.

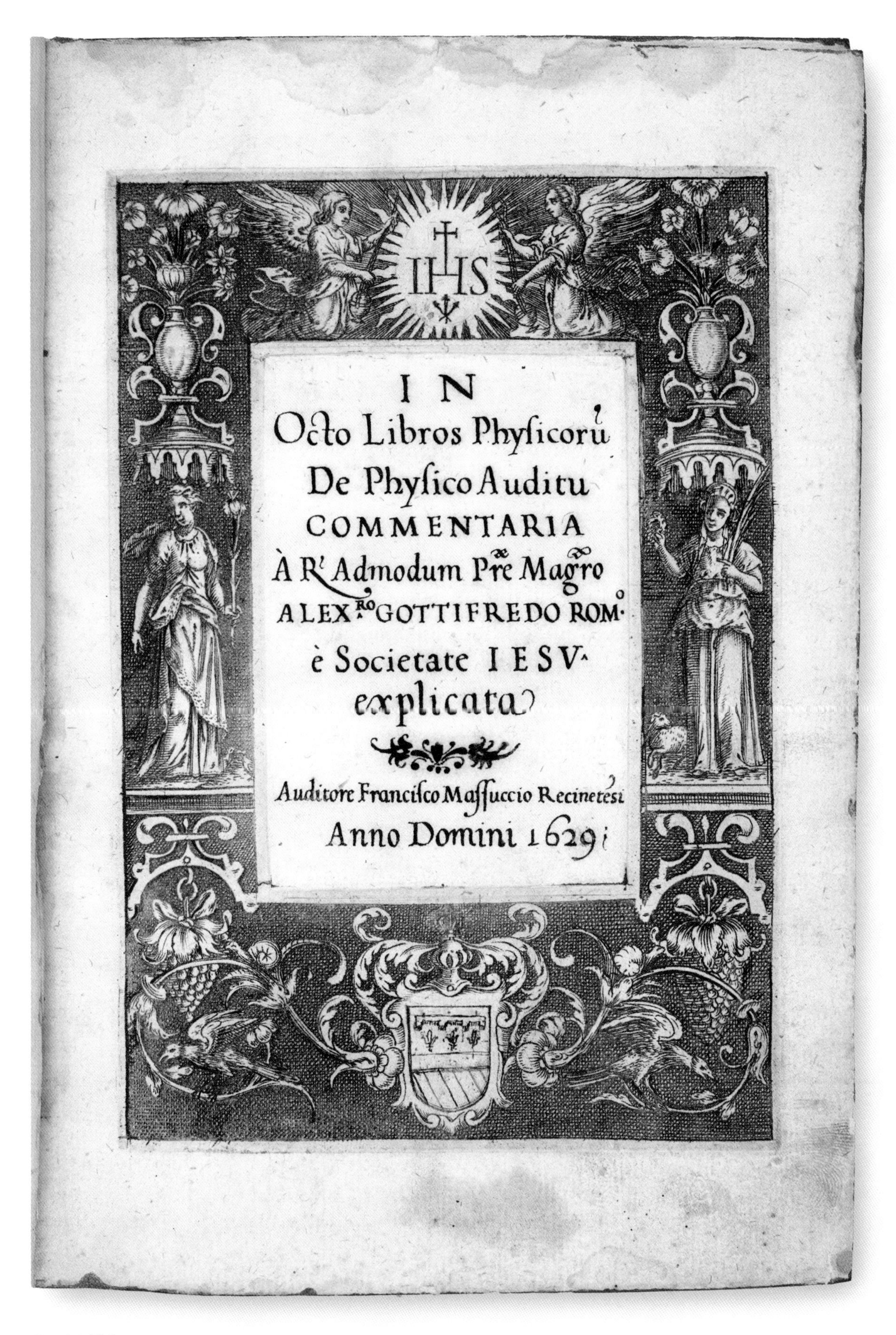
IHS

IN
Octo Libros Physicorũ
De Physico Auditu
COMMENTARIA
À R. Admodum Pre Magro
ALEX.ro GOTTIFREDO ROM.o
è Societate IESV
explicata

Auditore Francisco Massuccio Recinetesi
Anno Domini 1629.

Cat. 24. Title page

25

REVISING TRADITION

Cesare Cremonini (1550–1631)

Disputatio ultima; De coelo et mundo

Italy, c. 1630
Manuscript on paper

Cesare Cremonini was one of the most committed adversaries of Galileo (see cat. 21). At Padua he wrote extensively on Aristotelian matters, becoming one of the most famous and best-paid (at double the salary of Galileo) professors of natural philosophy. So popular was he that in his sixty-year career several kings and princes owned his portrait and corresponded with him. In 1613 he published his *Disputatio de Coelo*, reasserting Aristotle's picture of the universe to counter that of Galileo. This manuscript appears to be a revision of that treatise and thus an item of some interest. It was written by a disciple or a scribe, not long before the end of Cremonini's life. It represents an unexplored moment in the shift from pure Aristotelian orthodoxy to an acceptance of the discoveries brought to light by the new experimental science. Does this manuscript reveal a late shift on Cremonini's part to be more open to Galileo's conclusions, or does it reflect a disciple's efforts to continue his master's herculean work of harmonizing Aristotelian theory with the data amassed by the new science?

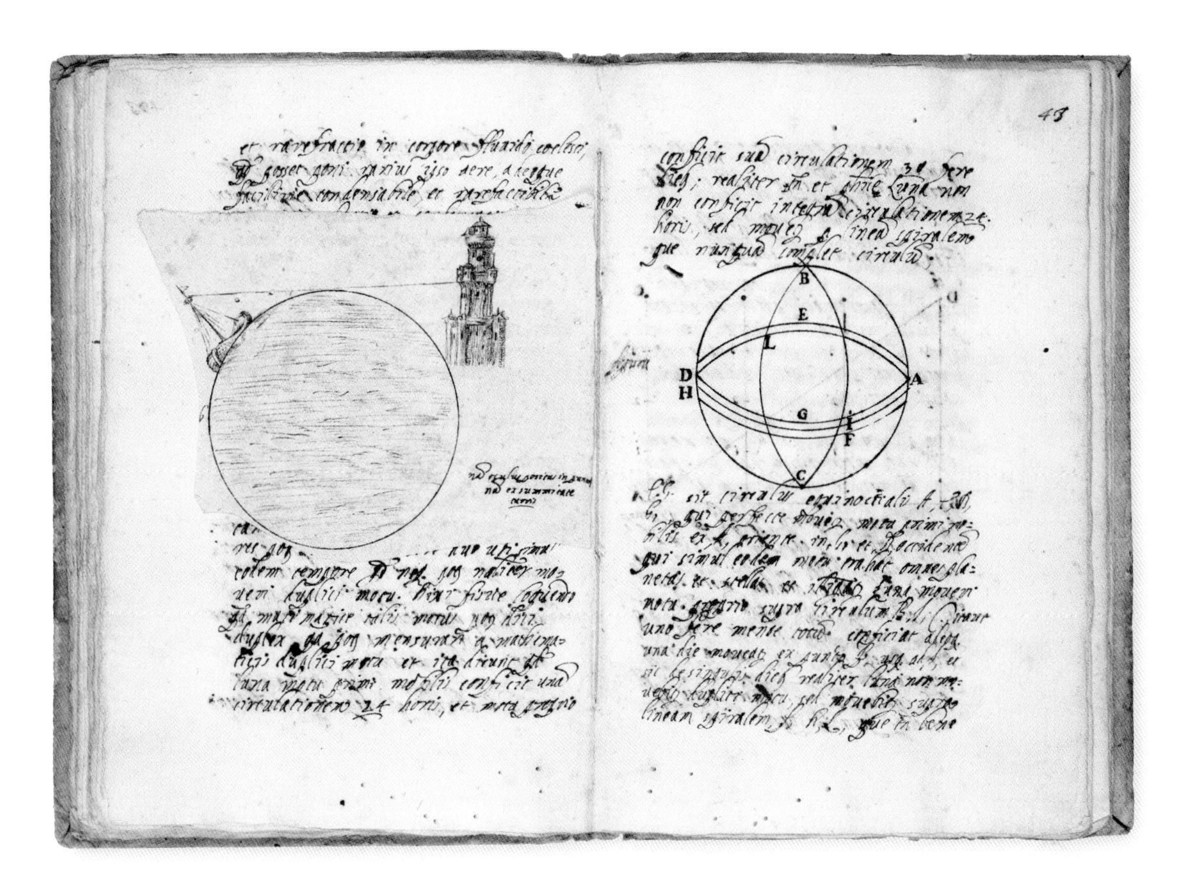

Cat. 25. Natural movement of the elements

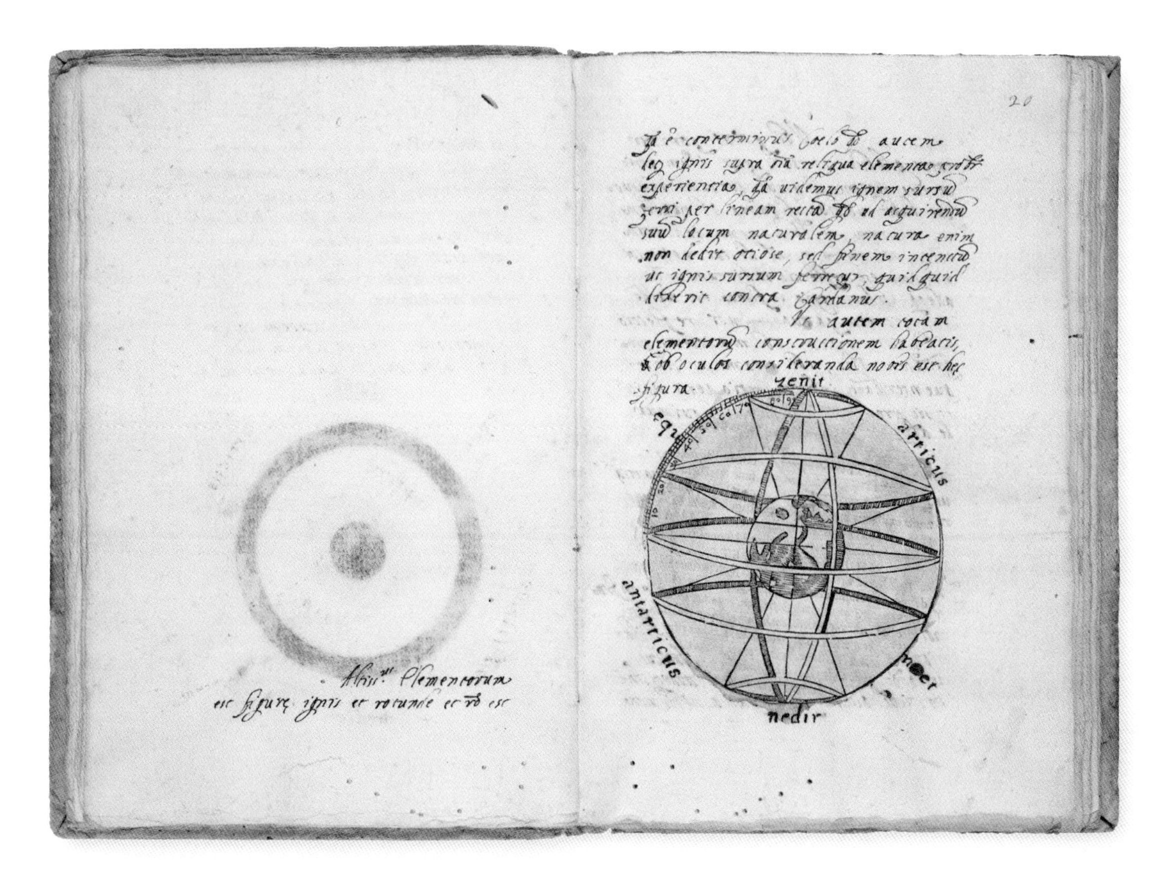

Cat. 25. Earth within an armillary sphere

26

A FRENCH PROTESTANT COMMENTARY

Anonymous

Commentarius in libros Phisicorum Aristotelis

France, c. 1650
Manuscript on paper

This anonymous work is in the form of an imaginary conversation between the author and the reader. It is written in Latin by a French scholar and consists of a detailed examination of Aristotelian natural philosophy. The composition probably dates from the second half of the seventeenth century, suggested both by the handwriting and by the frequent references to Descartes. The author includes Descartes, who died in 1650, in his historical overview as the most recent philosopher to have contributed to the progress of physics; his controversial positions were broadly disseminated, discussed, and debated, particularly in France.

Who was our anonymous author? It is likely that he was a French scholar familiar with the scientific controversies of the day. He regularly criticizes Descartes, but not from a position of conservative orthodoxy. In fact, some of his remarks suggest that he may have harbored Calvinist sympathies. Calvin's Protestant writings were anathema throughout Catholic Europe, and this may explain why the author chose to remain anonymous. The possibility that the author was a Calvinist is suggested by a detail in the binding: the arms are similar to those of the Derazes family from Poitou, a region in France that was a hotbed of Huguenot (French Calvinist) activity. The book was later owned by the Englishman John Towneley, fellow of the Royal Society and trustee of the British Museum, and by Emmanuel Louis Nicolas Viollet-le-Duc, a literary scholar and librarian at the Tuileries Palace in Paris.

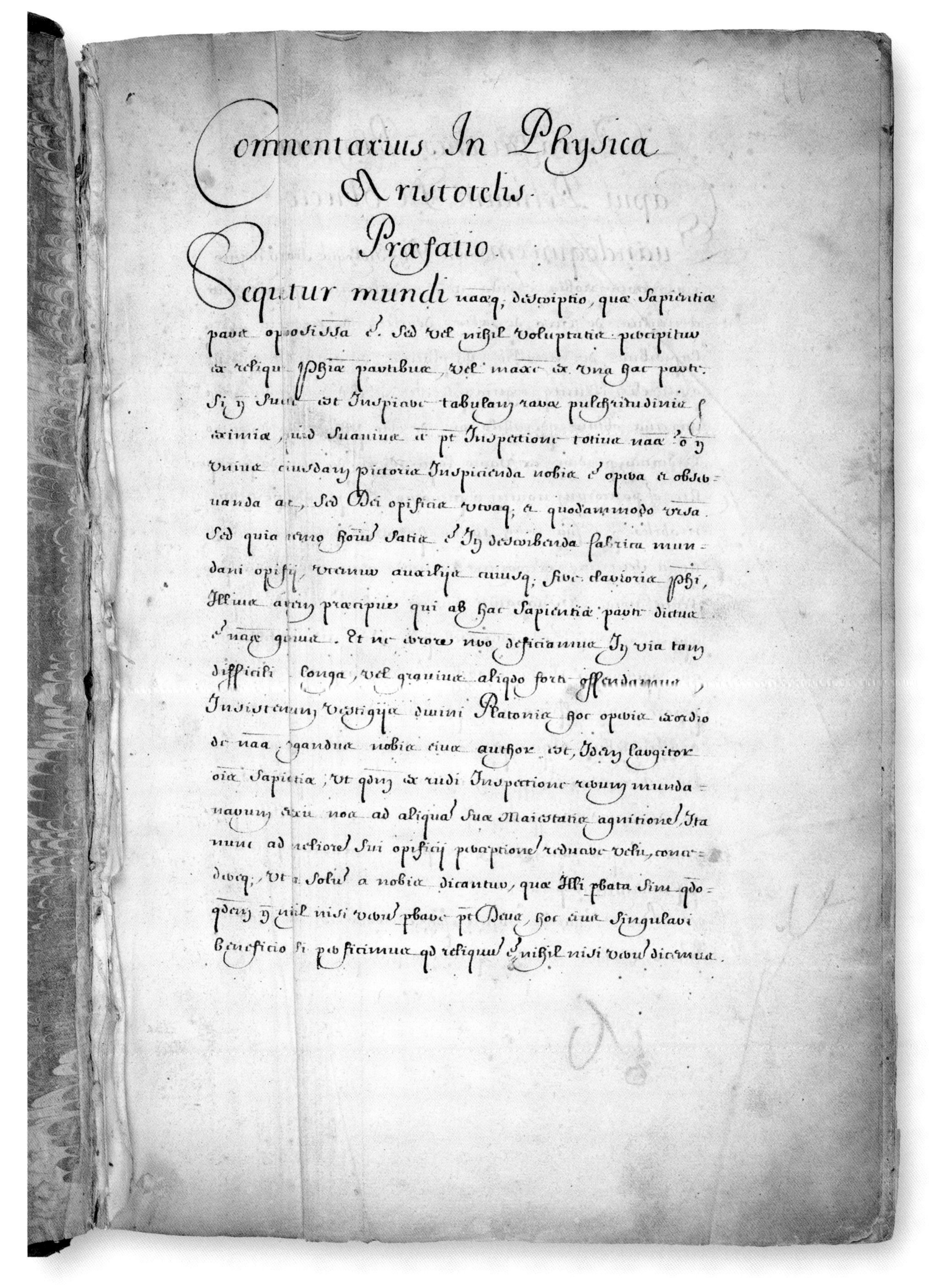

Commentarius In Physica Aristotelis.

Praefatio

Sequitur mundi naaq; descriptio, quae Sapientiae
pars operosissa est. Sed vel nihil voluptatis percipitur
ex reliquis Philosophiae partibus, vel maxe ex una hac parte.
Si n. suave est inspicere tabulam rarae pulchritudinis &
eximiae, quid suavius est quam inspectione totius naturae?
vivae eiusdem pictoris inspicienda nobis est opera et obser-
vanda est, sed Dei opificis utraque et quodammodo versa.
Sed quia unus homo satis est in describenda fabrica mun-
dani opificij, utamur auxilijs cuiusque sive clarioris Philosophi,
Illius autem praecipue qui ab hac Sapientia parte dictus
est naturae genius. Et ne errore nostro deficiamus in via tam
difficili longa, vel gravius aliquo forte offendamus,
Insistemus vestigijs divini Platonis hoc operis exordio
de natura agendus nobis eius author est, Deum largitor
omnis Sapientiae, ut quem ex rudi inspectione rerum munda-
narum excitavimus ad aliquam suae Maiestatis agnitionem, Ita
nunc ad meliorem sui opificij perceptionem reducere velit, conce-
datque, ut ea solum a nobis dicantur, quae illi probata sint quo-
quidem in nihil nisi verum probare potest Deus, hoc eius singulari
beneficio si perficimus quod reliquum est nihil nisi verum dicamus.

Cat. 26. Preface

27
MONASTIC LECTURE NOTES

Columbanus Parent (17th century)
and Antonius Chassé (17th century)

Octo libri Physicorum ad mentem D. Thomae; Commentarius in duos Aristotelis libros De generatione et corruptione; Tractatus in Metaphysicam Aristotelis; Commentarius in quatuor Aristotelis libros De coelo; Disputationes in tres libros Aristotelis De anima

Saint-Amand Abbey, France, 1664–65
Manuscript on paper

Abbeys, monasteries, and convents were for hundreds of years the places where knowledge was preserved and handed down. We owe our knowledge of many texts from antiquity to the patient work of monastic scribes who copied texts in their scriptoria. The rise of the universities in the thirteenth century and the invention of printing with moveable type in the later fifteenth century changed both the methods and the speed of the transmission of texts, but their work was based on the foundations provided by monastic scriptoria. Monasteries continued to play an important role in the education of elites in the sixteenth and seventeenth centuries. The present manuscript contains lecture notes for courses on Aristotle's natural sciences works and *Metaphysics* taught by the Benedictine professors Columban (Columbanus) Parent and Antoine (Antonius) Chassé in the novitiate school of Saint-Amand Abbey (formerly Elnon Abbey) in France.

The course in Aristotelian physics at the Benedictine abbey can be compared with the teaching in the universities of Louvain (Catholic) and Leiden (reformed), from which we have course records from the period. Isolated from the scrutiny that attached to university teaching, scholars at the abbey might have been more at liberty to investigate the implications of modern science. Thus, inserted into the manuscript is a document describing three contending world systems: those of Ptolemy, Tycho Brahe, and Copernicus. The lectures openly discuss the corruptible nature of the heavenly (as opposed to sublunary) bodies; they accept the new discoveries in optics that allowed scientists to explore more deeply man, nature, and the cosmos. But as bold as the lectures can seem, they do not quite go as far as embracing the heliocentric universe of Copernicus, remaining faithful to Aristotle and to the Church that supported him.

Cat. 27. Comparison of the systems of Ptolemy, Brahe, and Copernicus

Cat. 27. Astrolabe

28

AN ENGLISH CATHOLIC ON THE CONTINENT

John Manners (1609–1695)

Disputationes in octo libros Physicorum

Perugia, Italy, 1647
Manuscript on paper

The life of John Manners (also known as John Simcocks, and later John Grosvenor) illustrates the itinerant, European destiny of those English intellectuals who remained within the Catholic Church when to do so meant ostracism in England. Manners was born in London and educated at the English College at Saint-Omer in Flanders, before entering the Society of

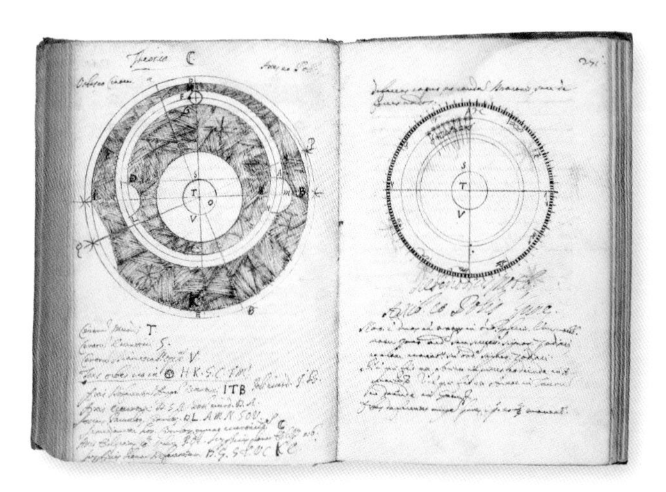

Cat. 28. How to calculate the position of a planet using Ptolemy's epicycles

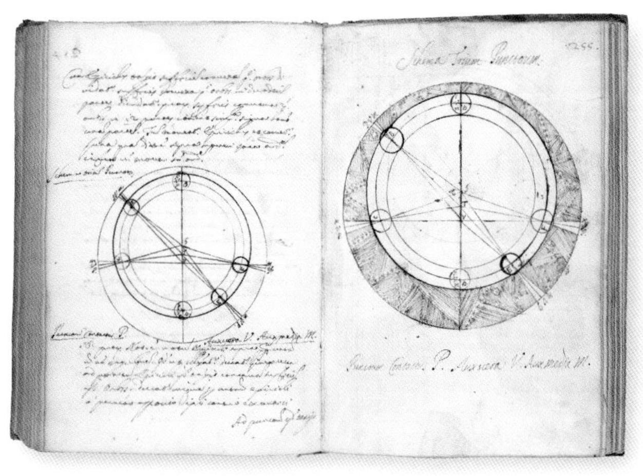

Cat. 28. "Diagram of the three points" for the reading of an astrolabe

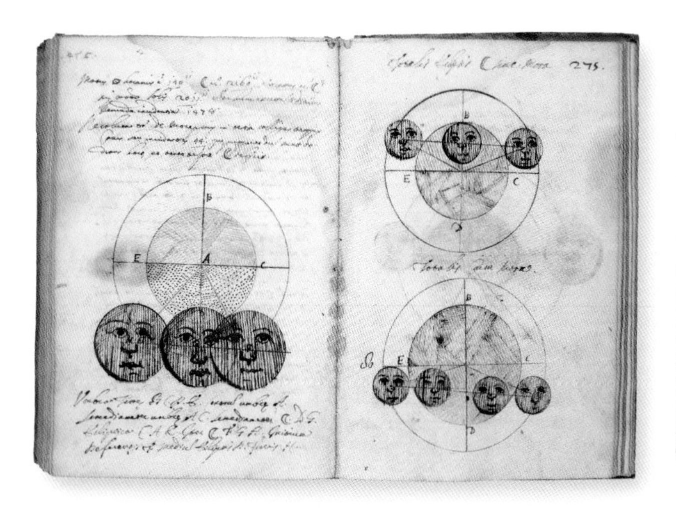

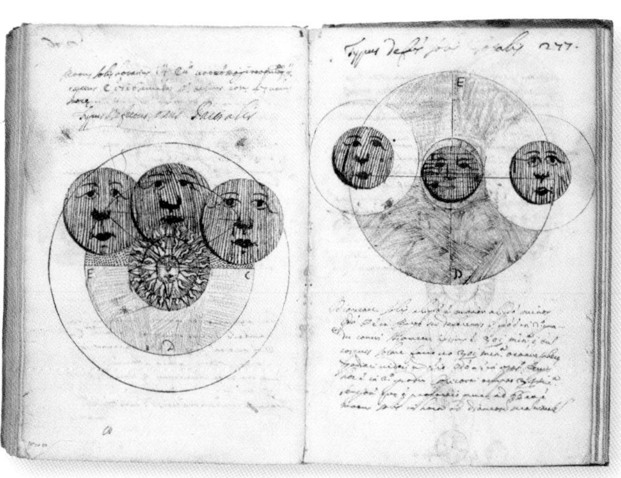

Cat. 28. Phases of the moon

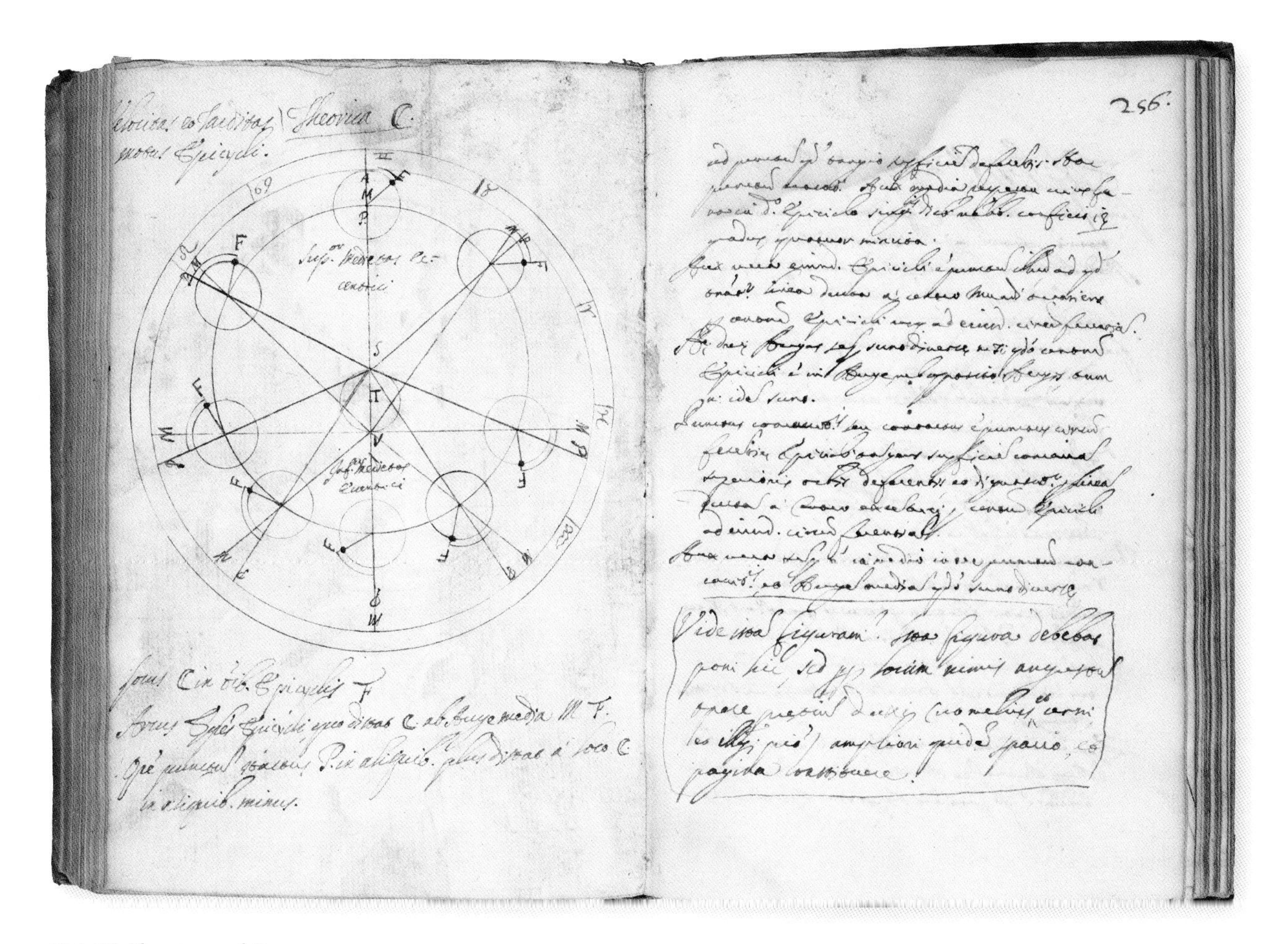

Cat. 28. Planetary orbits

Jesus. He began lecturing in philosophy at the University of Perugia following his ordination in 1640. He taught there while the English Civil War was raging, and read the lectures in the same year that Parliament abolished Christmas. His student, Franciscus Pavetius of Cantiano, is credited as the scribe of this manuscript, including the fifty-five diagrams illustrating all aspects of physics. By 1649, Manners must have proved his worth: he moved to the English College at Rome, where he served as director of studies and rector. After a time in Liège, he successfully made a return to British soil (Suffolk and London). He later retired to France, staying at Saint-Germain and the court of the exiled Catholic king James II until his death.

29
BEING BOLD IN SMALL WAYS

Giovanni Carlo Maffei (17th century)

In quatuor libri de coelo ad subtilium principis mentem

Northern Italy, 1677
Manuscript on paper

This unassuming manuscript brings us to a defining moment in the history of Western thought. Understanding its content as well as its appearance requires a little context. Aristotle's view of the universe, with Earth at its core, surrounded by unchanging concentric spheres, each hosting celestial bodies, was engrained in Western thought for nearly two millennia. For a very long time, philosophers and members of the Christian clergy were largely in agreement that the Earth was at the center of the universe, the heavens above were immutable, and man was the ultimate center. Then in 1543 the Polish scientist Copernicus (1473–1543) gave a revolutionary account: the sun, in fact, was at the center of our universe; the Earth was merely one of a number of orbiting satellites. This, philosophically, amounted to the shattering assertion that mankind, no longer easily enthroned as powerful sovereign with God's endorsement, was but a mere inhabitant of a peripheral body.

Copernicus's theory did not initially cause much scandal within the Catholic Church; the lack of tangible, irrefutable evidence meant that it could remain just that—a theory, mainly for the consumption of scholars. Then in 1573 Tycho Brahe (1546–1601), who had methodically observed the skies with sophisticated tools, published his account of the appearance of a new bright star, a moving star—known today as the supernova SN 1572. The supernova challenged the fixity of the heavens. How could a mass travel freely across distances and depths in the sky if the universe was really made of crystal-hard, unmovable, unchanging concentric celestial spheres? What is more, Brahe showed that unassisted sensory perception, relied upon since Aristotle's time for the building of knowledge, could be misleading: the discovery of truth required evidence, and evidence was to be obtained through new, well-calibrated instruments. Galileo's observations, conducted in this vein, all but confirmed the new science. And so it was only in 1616, and then again in 1633, that the work of Copernicus was subjected to a ban by the Church—together with Galileo's. It was more than a century later, in 1758, that the church finally dropped its prohibition against books promoting heliocentrism.

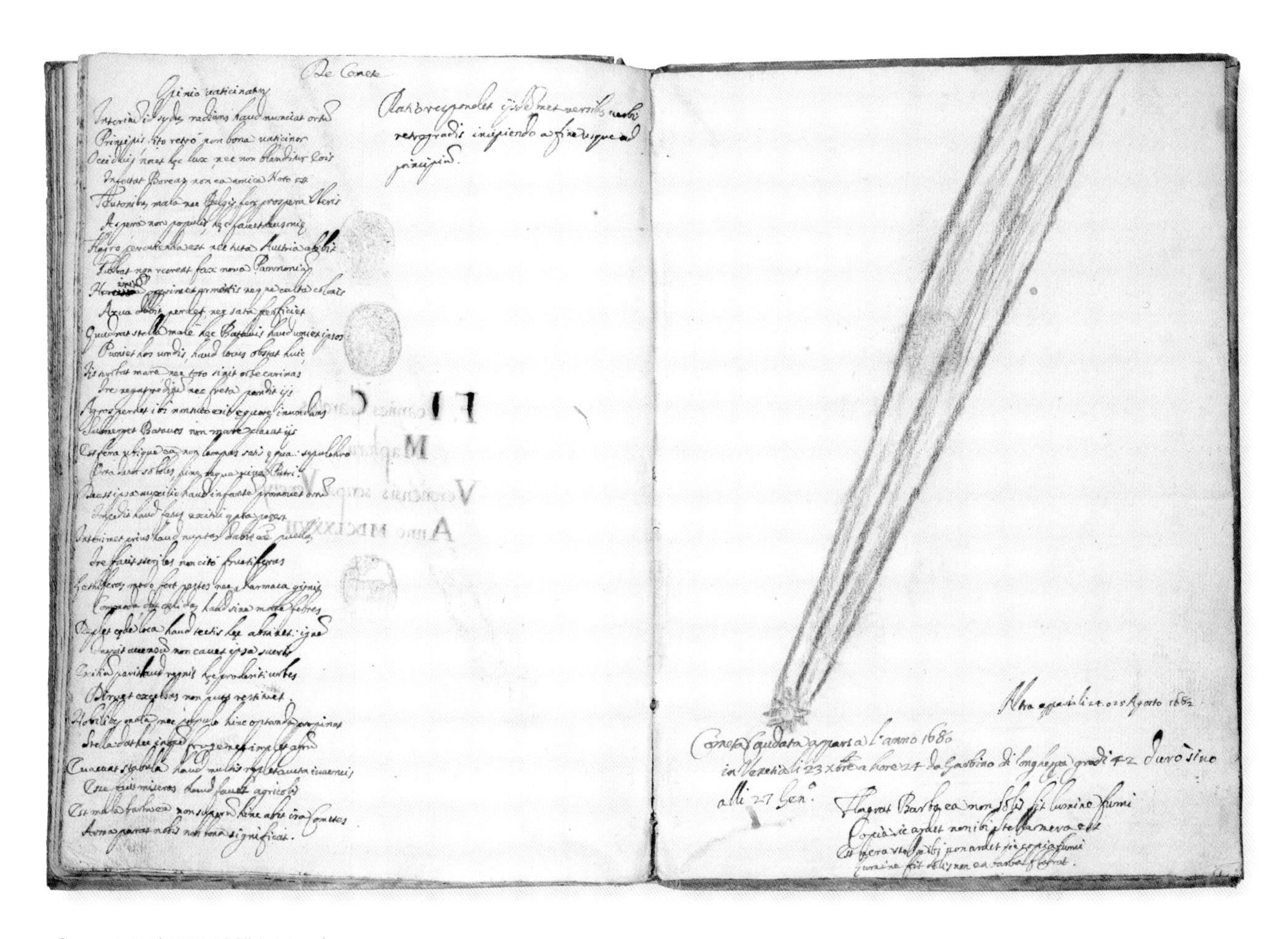

Cat. 29. Sighting of the comet

The author of this striking notebook, Giovanni Carlo Maffei, a largely unknown seventeenth-century scholar from northern Italy, seeks to grapple with the new cosmology. It was a risk to start from controversial premises and then to try to square them with a millennium of received science and philosophy that included notions of morals and of metaphysics. His manuscript included an account of the Great Comet of 1680, a phenomenon that again subverted the fixity of the celestial spheres. Maffei must have been aware that he could run afoul of the authorities. How could he carry on with his examination of Aristotle's thought without risking a charge of heresy? How could he protect his manuscript and himself from the curiosity of even a casual reader? Maffei had a solution: he redacted his commentary in microscript, readable only with the help of magnification, and cropped most words, filling the pages with barely readable abbreviations. Only a highly trained, highly educated investigator would find reason to be suspicious of this little notebook. This is a very private, deliberately cryptic document. Even in late-seventeenth-century Italy, the new science had to wrap itself in stealth.

30

MODERNIZING ARISTOTLE

Aristotle

De natura; De caelo; De mundo; De ortu et interitu; Meteorologicorum libri; De anima; Parva naturalia

Paris: Gabriel Buon, 1560
Annotated by a student of James Martin (16th century)

The Scottish Renaissance philosopher James Martin (Jacobus Martinus, Jacques Martin) lived in the sixteenth century, probably studied in Oxford, and later held university teaching positions in Paris and Turin. Martin was a modernizer and a critic of Aristotle. He found Aristotle's large and disparate body of work unsatisfying because it lacked a consistent methodological approach. Few records of Martin's life and work survive. The marginalia in this book, probably written by one of Martin's students, appear to be the only extant record of lectures on Aristotle he delivered in Paris in 1563. It was in such lectures that Martin worked out the details of his critique of Aristotle, which he saw into print in 1577.

LIBER II.

Forma autem vtrum principium vnum sit, an multa, & quodnam, vel quae sint, quoniam primae philosophiae est accurate explicare, ad illud tempus differatur. De formis vero quae naturales sunt & intereunt, deinceps differemus. Ac principia quidem esse, quaeque, ac quot sint numero, a nobis hactenus explicatum sit. Nunc aliunde exordiamur.

ARISTOTELIS DE NATVRA, AVT DE RERVM INITIIS LIBER II.

Cap. I.

Eorum quae sunt in natura rerum, alia natura sunt & constant, alia ab aliis causis originem ducunt. Natura, vt animantia, & eorum partes, & ea quae oriuntur e terra, & simplicia corpora: vt terra, ignis, aër, & aqua. Haec enim, quaeque sunt eiusdem generis, natura constare dicimus. Omnia autem quae enumerauimus, hoc ab iis quae naturae ordine orta non sunt, differre constat, quod ea quae natura constat, intra semetipsa motus & status principia habere videtur, partim in loco, partim crescendo & senescendo, alia commutando. lectus vero & tunica, & si quod est aliud tale genus, quatenus suo quodque nomine appellatur, & qua ex parte artis sunt, nullum mutationis insitum principium habent: quatenus autem eis contigit, vt ex lapide, vel terra sint, vel ex horum permixtione, eatenus habent: quasi

Cat. 30. *On the Nature of Things*, book II, with notes from Martin's lectures

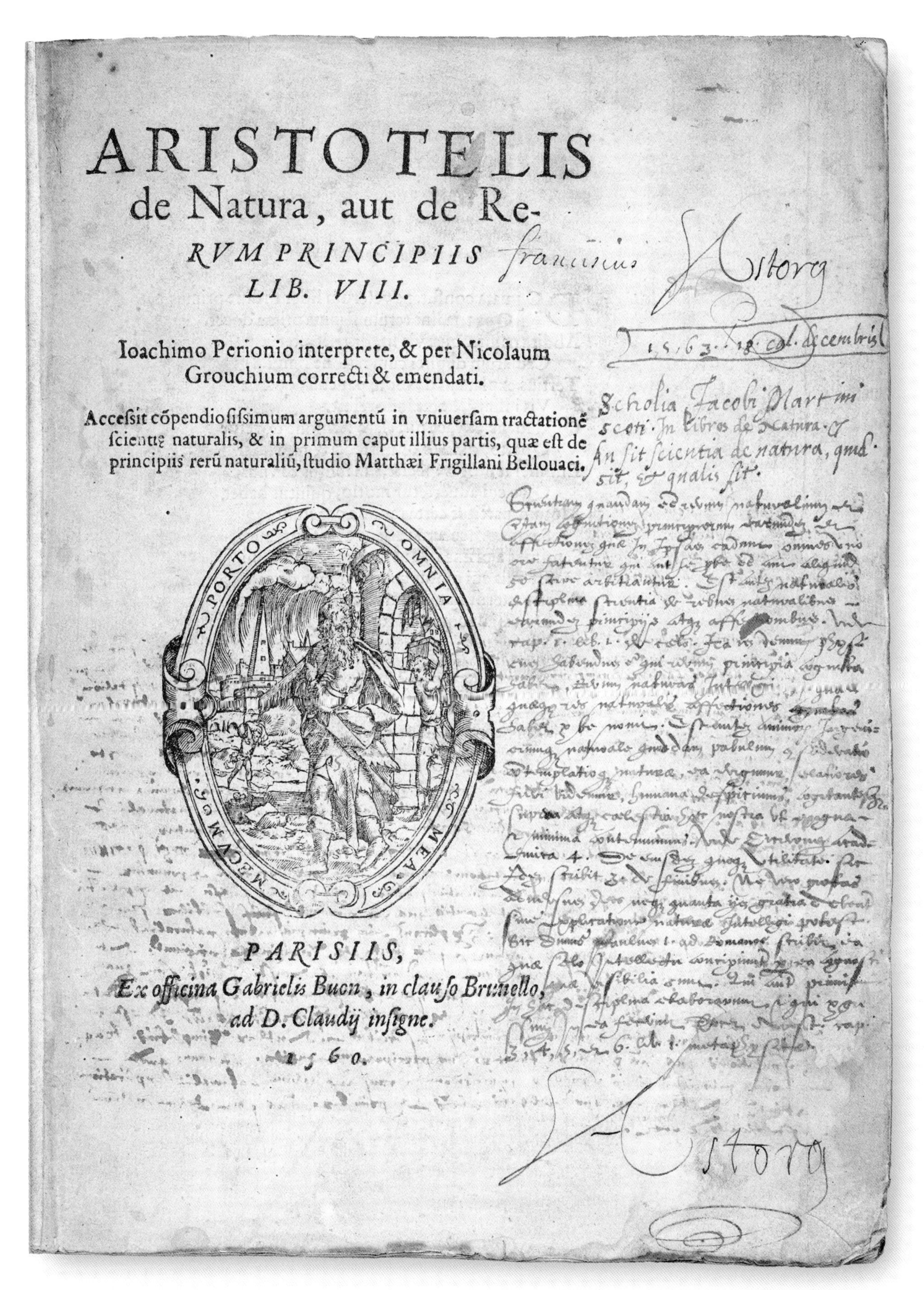

ARISTOTELIS
de Natura, aut de Re-
RVM PRINCIPIIS
LIB. VIII.

Ioachimo Perionio interprete, & per Nicolaum Grouchium correcti & emendati.

Accesit cõpendiosissimum argumentũ in vniuersam tractationẽ scientiæ naturalis, & in primum caput illius partis, quæ est de principiis rerũ naturaliũ, studio Matthæi Frigillani Bellouaci.

PARISIIS,

Ex officina Gabrielis Buon, in clauso Brunello, ad D. Claudij insigne.

1560.

Cat. 30. Title page with annotations

Ethicorum Aristotelis

Opus Aristotelis de moribus ad Nicomachum: a Johanne argyropilo Byzantio: Causa clarissimi viri Cosme medicis Florentini traductum.

Tractatus Primus.

¶ Capite primo Aristoteles ostendit omnia bonũ expetere: diuersos esse rerum fines: sed semper extremũ optimũ esse. ceterisqꝫ prestare.

Omnis ars: omnisqꝫ doctrina atqꝫ actus: itidem et electio / bonum quoddam appetere videtur. Quapropter bene veteres bonũ ipsum id esse dixerunt quod omnia appetunt. ¶ Finium autem differentia quedã esse videtur. Quidam enim sunt operationes: quidam preter has ipsas opera quedam. Atqꝫ in quibus preter operatiões. alij quidã sunt fines / in his ipsa opera sunt operationibus prestabiliora. ¶ Cum vero complures sint actus et artes. atqꝫ scientie / fit vt multi sint etiã fines. Medicine nãqꝫ finis est sanitas. Artis extruendarũ nauium. nauis: facultatis rei militaris. victoria: et familiaris rei gubernande / diuitie. Que autem talium sub aliqua vna potentia sunt (vt ars cõficiendorũ frenorũ. ceterecꝫ omes instrumentorũ equestriũ effectrices / sese habent ad facultatẽ equestrẽ: et hec cum illis. omnisqꝫ actio bellica ad militarem facultatẽ: et ad alias alie simili modo) in his omnib9. fines earũ que rationẽ architecture subeũt / magis sunt expetibiles quam oẽs fines inferiorũ: Et illi em̃ horũ gratia expetuntur. Atqꝫ nihil interest. siue fines actuũ ipse sint operationes: siue preter ipsas aliquid aliud vt in dictis facultatibus intueri licet.

¶ Cap. ij. Ex ratione vltimi finis / summũ bonũ humanum inquirit: cui9 cognitio ad vitã humanã multũ cõfert: docet sciẽtiã hãc. que de illo est / esse ciuilẽ: id est. q̃ ad ciuiũ mores. honestatẽ. cõmoditatem spectet.

ETHICS AND POLITICS

"The wise person, more than anyone else, will be happy."

NICOMACHEAN ETHICS, BOOK X, CHAPTER 8

How do we live a good life? What makes us happy; what stands in the way? What is justice, and how do we tell a fair transaction from an unjust one? How do we articulate public life so that it is oriented towards both prosperity and justice? Can a ruler be above the law? Is money an effective way of comparing the value of non comparable goods? What is value? No corner of human interaction was left unexplored by Aristotle, who could not conceive of mankind without taking into account man's innate sociability. Readers of Aristotle have engaged with these questions throughout the centuries.

Aristotle's belief in the cultivation of virtues and the building of a good character through good habits was wholly endorsed by Christian theology in the Middle Ages. His discussion of the essence of value shaped Thomas Aquinas's theory of "just price" and later generated the pioneering economic thought at the University of Salamanca in Spain. His focus on happiness as the ultimate goal in life invigorated the Renaissance shift from duty to personal fulfillment; his emphasis on rational choice shaped discussions of free will as Reformation thinkers strove to foreground the role of divine grace; and his insistence upon the rule of law animated early modern reflections on the limits to be imposed on rulers.

Aristotle and John Argyropoulos (1415–1487)
Opus ... de moribus ad Nicomachum ...
a Joannae Argyropilo ... traductum
Leipzig: Martin [Landsberg], 1511
(detail of cat. 35)

31

THE CLASSIC TRANSLATION OF THE ETHICS

Aristotle and Robert Grosseteste (translator, 1175–1253)

Ethica ad Nicomacum

Northern Italy, c. 1425
Manuscript on paper

Aristotle's *Nicomachean Ethics* is his best-known work on ethics. The underlying question addressed in this treatise is how to live a good life. What is happiness? What are the virtues that make it possible? This belief in an ethics of virtue and in the formation of a good character through good habits fits well with medieval Christianity. Robert Grosseteste, first chancellor of Oxford University and bishop of Lincoln, was a towering figure in the fields of politics, science, philosophy, and theology. He must have learned Greek late in life, possibly as a bishop—a position that allowed him to obtain Greek manuscripts. His translation of Aristotle's *Nicomachean Ethics*, thought to have been accomplished around 1247 and known in the Middle Ages as the *Liber Ethicorum*, was essential for medieval ethical thought. This manuscript contains a complete copy of Grosseteste's Latin translation of the *Nicomachean Ethics*, in a very early and now rare version preceding the revisions made at a later stage by Grosseteste himself or by another great editor, William of Moerbeke. It was clearly produced in a center of high culture: the script is an elegant gothico-antiqua; the larger decorated initials point to humanist centers in northern Italy, such as Padua. The sheets are bound with manuscript fragments quoting the work of various classical authors, including Terence and Plautus.

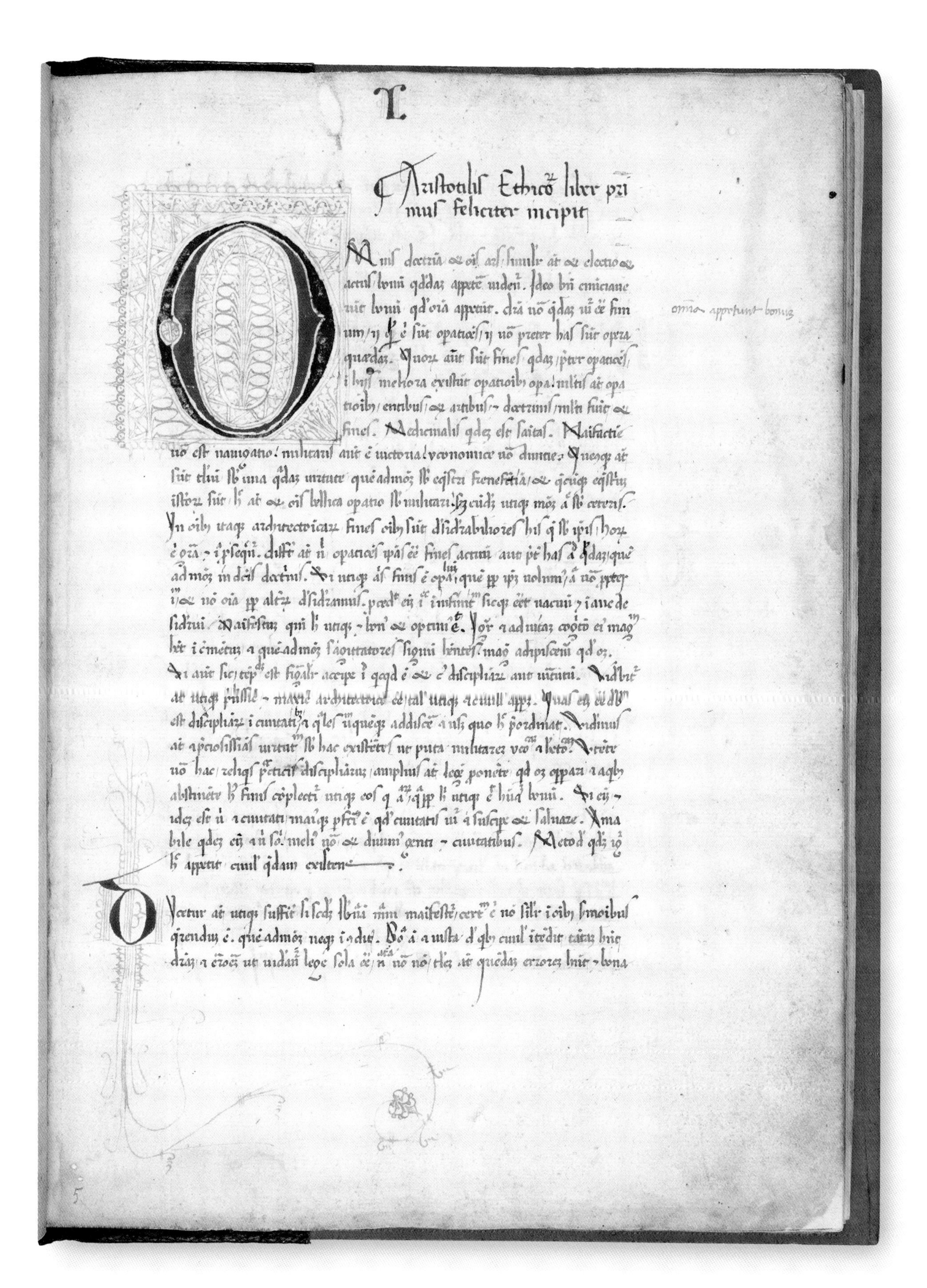

Cat. 31. First page

32
PRINTING THE ETHICS

Robert Grosseteste (1175–1253) **and John Buridan** (c. 1300–c. 1358)

Decem librorum Moralium Aristotelis

Paris: Johannes Higman and Wolfgang Hopyl, 12 April 1497

A pillar of medieval ethical thought, Grosseteste's translation of Aristotle's *Nicomachean Ethics* (cat. 31) continued to be read well into the Renaissance. Grosseteste liked to translate the text literally; he would add notes clarifying especially obscure passages when the Latin was not as self-explanatory as the Greek. This made his translation, complemented by William of Moerbeke's revisions, a standard text even in the age of the printing press, when a variety of competing versions became more widely available. This edition also contains the commentary of John Buridan (c. 1300–c. 1358), whose position on the topic of free will is based on Aristotle's *Nicomachean Ethics*. Buridan's name is often associated with "Buridan's Ass," a thought experiment in which an imaginary donkey stands midway between two equal stacks of hay; unable to choose which bale to eat, the donkey starves. The figure of Buridan's ass satirizes Buridan's belief that faced with two equally compelling alternatives, we must wait to choose until conditions change and one option becomes preferable to the other.

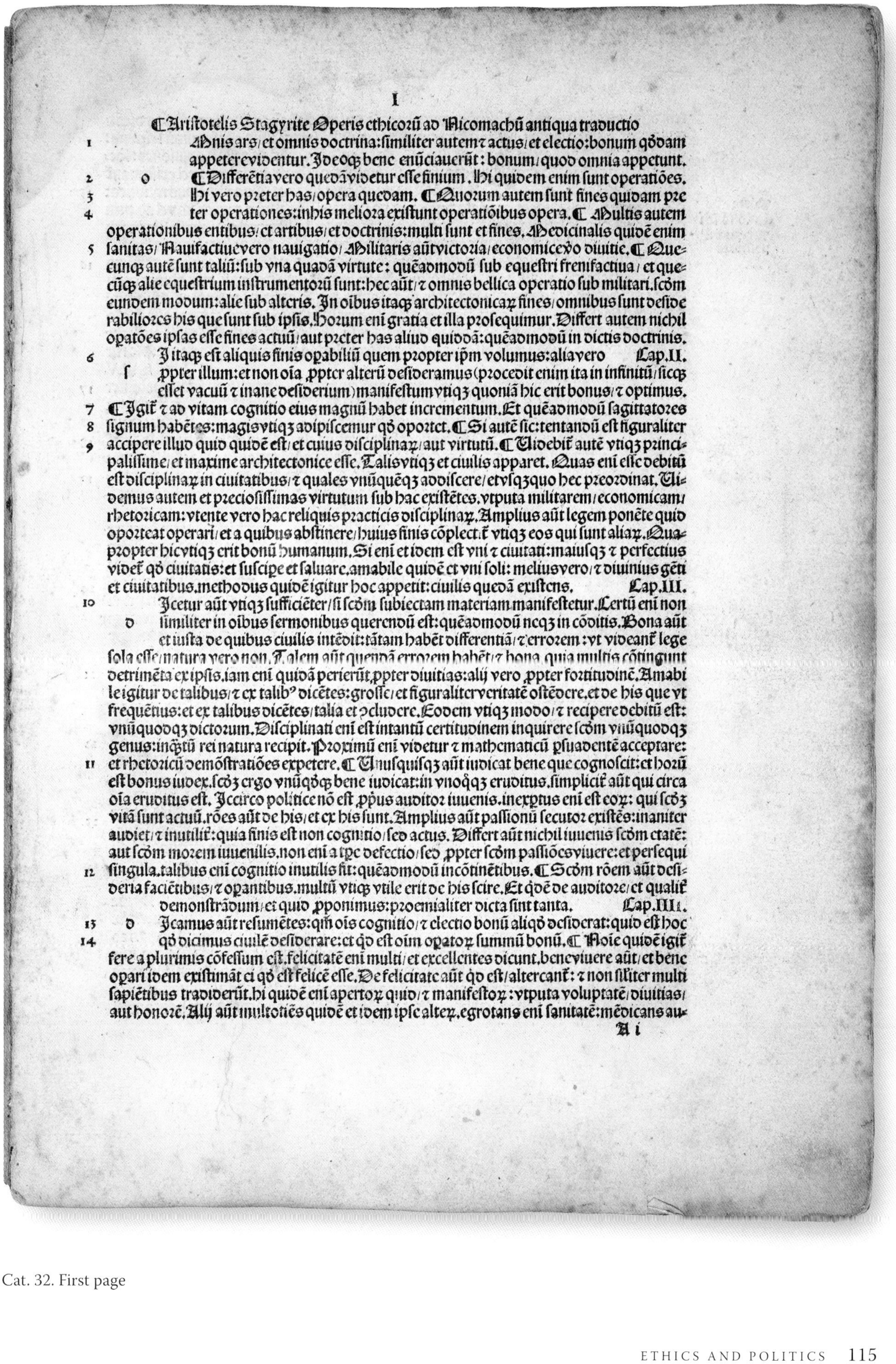

I

¶ Aristotelis Stagyrite Operis ethicorū ad Nicomachū antiqua traductio

1 Omnis ars, et omnis doctrina: similiter autem ⁊ actus, et electio: bonum qddam
appetere videntur. Ideoqꝫ bene enūciauerūt: bonum, quod omnia appetunt.
2 o ¶ Differētia vero quedā videtur esse finium. Hi quidem enim sunt operatiões.
3 Hi vero preter has, opera quedam. ¶ Quorum autem sunt fines quidam pre
4 ter operationes: in his meliora existunt operatiōibus opera. ¶ Multis autem
operationibus entibus, et artibus, et doctrinis: multi sunt et fines. Medicinalis quidē enim
5 sanitas, Nauifactiue vero nauigatio, Militaris aūt victoria, economice vō diuitie. ¶ Que-
cunqꝫ autē sunt taliū: sub vna quadā virtute: quēadmodū sub equestri frenifactiua, et que-
cūqꝫ alie equestrium instrumentorū sunt: hec aūt, ⁊ omnis bellica operatio sub militari. scdm
eundem modum: alie sub alteris. In oībus itaqꝫ architectonicaꝝ fines, omnibus sunt deside
rabiliores his que sunt sub ipsis. Horum enī gratia et illa prosequimur. Differt autem nichil
opatōes ipsas esse fines actuū, aut preter has aliud quiddā: quēadmodū in dictis doctrinis.
6 Si itaqꝫ est aliquis finis opabiliū quem propter ipm volumus: alia vero Cap. II.
s ppter illum: et non oīa ppter alterū desideramus (procedit enim ita in infinitū, sicqꝫ
esset vacuū ⁊ inane desiderium) manifestum vtiqꝫ quoniā hic erit bonus, ⁊ optimus.
7 ¶ Igit ⁊ ad vitam cognitio eius magnū habet incrementum. Et quēadmodū sagittatores
8 signum habētes: magis vtiqꝫ adipiscemur qd oportet. ¶ Si autē sic: tentandū est figuraliter
9 accipere illud quid quidē est, et cuius disciplinaꝝ, aut virtutū. ¶ Videbit autē vtiqꝫ princi-
palissime, et maxime architectonice esse. Talis vtiqꝫ et ciuilis apparet. Quas enī esse debitū
est disciplinaꝝ in ciuitatibus, ⁊ quales vnūquēqꝫ addiscere, et vsqꝫ quo hec preordinat. Vi-
demus autem et preciosissimas virtutum sub hac existētes. vtputa militarem, economicam,
rhetoricam: vtente vero hac reliquis practicis disciplinaꝝ. Amplius aūt legem ponēte quid
oporteat operari, et a quibus abstinere, huius finis cōplect vtiqꝫ eos qui sunt aliaꝝ. Quapropter hic vtiqꝫ erit bonū humanum. Si enī et idem est vni ⁊ ciuitati: maiusqꝫ ⁊ perfectius
videt qd ciuitatis: et suscipe et saluare. amabile quidē et vni soli: melius vero, ⁊ diuinius gēti
et ciuitatibus. methodus quidē igitur hoc appetit: ciuilis quedā existens. Cap. III.
10 Dicetur aūt vtiqꝫ sufficiēter, si scdm subiectam materiam manifestetur. Certū enī non
d similiter in oībus sermonibus querendū est: quēadmodū neqꝫ in cōditis. Bona aūt
et iusta de quibus ciuilis intēdit: tātam habēt differentiā, ⁊ errorem: vt videant lege
sola esse, natura vero non. Talem aūt quendā errorem habēt, ⁊ bona. quia multis cōtingunt
detrimēta ex ipsis. iam enī quidā perierūt ppter diuitias: alij vero ppter fortitudinē. Amabi
le igitur de talibus, ⁊ ex talibꝰ dicētes: grosse, et figuraliter veritatē ostēdere. et de his que vt
frequētius: et ex talibus dicētes, talia et ꝯcludere. Eodem vtiqꝫ modo, ⁊ recipere debitū est:
vnūquodqꝫ dictorum. Disciplinati enī est intantū certitudinem inquirere scdm vnūquodqꝫ
genus: inqtū rei natura recipit. Proximū enī videtur ⁊ mathematicū psuadentē acceptare:
11 et rhetoricū demōstratiōes expetere. ¶ Vnusquisqꝫ aūt iudicat bene que cognoscit: et horū
est bonus iudex. scdꝫ ergo vnūqdqꝫ bene iudicat: in vnoqqꝫ eruditus. simplicit aūt qui circa
oīa eruditus est. Iccirco politice nō est ppus auditor iuuenis. inexptus enī est eoꝝ: qui scdꝫ
vitā sunt actuū. rōes aūt de his, et ex his sunt. Amplius aūt passionū secutor existēs: inaniter
audiet, ⁊ inutilit: quia finis est non cognitio, sed actus. Differt aūt nichil iuuenis scdm etatē:
aut scdm morem iuuenilis. non enī a tpe defectio, sed ppter scdm passiōes viuere: et persequi
12 singula. talibus enī cognitio inutilis fit: quēadmodū incōtinētibus. ¶ Scdm rōem aūt desi-
deria faciētibus, ⁊ opantibus. multū vtiqꝫ vtile erit de his scire. Et qdē de auditore, et qualit
demonstrādum, et quid pponimus: proemialiter dicta sint tanta. Cap. IIII.
13 d Dicamus aūt resumētes: qm oīs cognitio, ⁊ electio bonū aliqd desiderat: quid est hoc
14 qd dicimus ciuilē desiderare: et qd est oīm opatoꝝ summū bonū. ¶ Noīe quidē igit
fere a plurimis cōfessum est. felicitatē enī multi, et excellentes dicunt. bene viuere aūt, et bene
opari idem existimāt ei qd est felicē esse. De felicitate aūt qd est, altercant: ⁊ non silr multi
sapiētibus tradiderūt. hi quidē enī apertoꝝ quid, ⁊ manifestoꝝ: vtputa voluptatē, diuitias,
aut honorē. Alij aūt multotiēs quidē et idem ipse alteꝝ. egrotans enī sanitatē: mēdicans au-
A i

Cat. 32. First page

33

AQUINAS AND SCHOLASTICISM

Thomas Aquinas (1225–1274)

Summa theologiae (parts 2.1 and 3)

Venice: Andrea Torresani, 1483; Bernardino de Tridino, 1486

Thomas Aquinas's *Summa Theologiae* is one of the most influential works of Western philosophy. The English philosopher Anthony Kenny considers Aquinas to be "one of the dozen greatest philosophers of the western world."[1] An Italian Dominican philosopher, theologian, and jurist, Aquinas was responsible for the medieval reconciliation between reason and revelation. He believed that we attain knowledge from two sources: faith, which shows us the ultimate truths pertaining to the divine realm, and reason, which shows us natural truths that ancient philosophers such as Aristotle had systematized. The *Summa*, an ambitious project, was an encyclopedia of man's knowledge about God, the universe, nature, and man himself, his intellect, will, and passions. Deeply informed by Aristotle, the *Summa*'s logical rigor and broad scope ensured the work's place in the canon of Western philosophy. Some historians have likened the aspirations and grandeur of the *Summa* to the contemporary architecture of gothic cathedrals. The *Summa* is divided into three parts, each printed at different times by different publishers; no collected edition was published until 1485. The first part treats of the nature and attributes of God, including the physical universe; the third part, which was completed according to Aquinas's plan after his death, deals with Christ. The second part is devoted to man in society, and to ethics, the ultimate ends or goals of man.

NOTE

1. Anthony Kenny, ed., "Introduction," in *Aquinas: A Collection of Critical Essays* (London: Macmillan, 1969), 1.

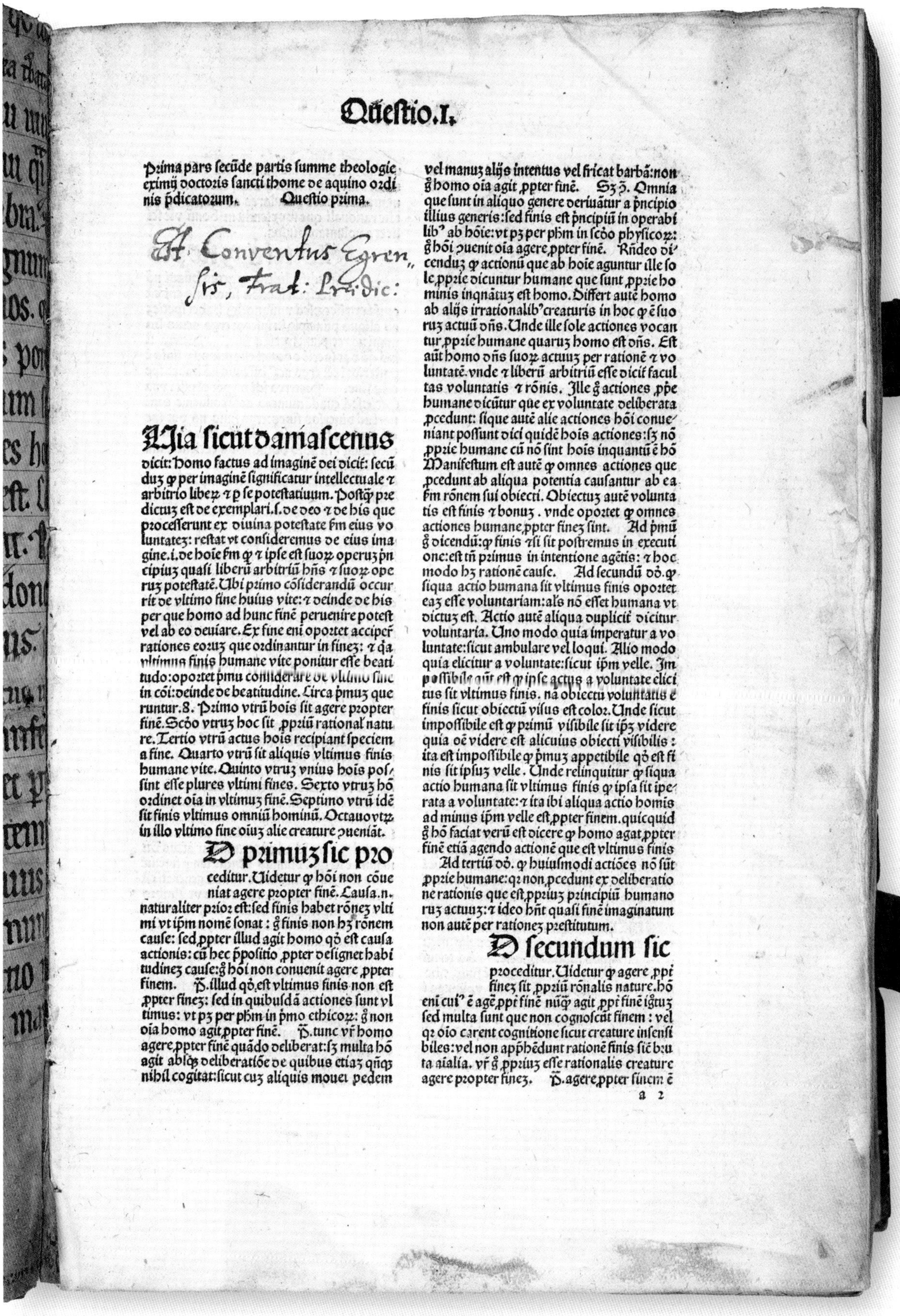

Questio.I.

Prima pars secūde partis summe theologie eximij doctoris sancti thome de aquino ordinis pdicatorum. Questio prima.

Ad Conventus Egrensis, Frat: Praedic:

Quia sicut damascenus dicit: homo factus ad imaginē dei dicit: secūdum q per imaginē significatur intellectuale et arbitrio liberum et p se potestatiuum. Postq̄ predictum est de exemplari. s. de deo et de his que processerunt ex diuina potestate fm eius voluntatem: restat vt consideremus de eius imagine. i. de hoie fm q et ipse est suorum operum pncipium quasi liberū arbitriū hñs et suorum operum potestatē. Ubi primo cōsiderandū occurrit de vltimo fine huius vite: et deinde de his per que homo ad hunc finē peruenire potest vel ab eo deuiare. Ex fine eni oportet accipere rationes eorum que ordinantur in finem: et quia vltimus finis humane vite ponitur esse beatitudo: oportet pmū considerare de vltimo fine in cōi: deinde de beatitudine. Circa pmum queruntur. 8. Primo vtrū hois sit agere propter finē. Scdo vtrum hoc sit pprium rational nature. Tertio vtrū actus hois recipiant speciem a fine. Quarto vtrū sit aliquis vltimus finis humane vite. Quinto vtrum vnius hois possint esse plures vltimi fines. Sexto vtrum hō ordinet oia in vltimum finē. Septimo vtrū idē sit finis vltimus omniū hominū. Octauo vtrum in illo vltimo fine oium alie creature conueniāt.

Ad primum sic proceditur. Uidetur q hōi non conueniat agere propter finē. Causa. n. naturaliter prior est: sed finis habet rōnem vltimi vt ipm nomē sonat: g finis non hz rōnem cause: sed ppter illud agit homo qd est causa actionis: cū hec ppositio ppter designet habitudinem cause: g hōi non conuenit agere ppter finem. Pr. illud qd est vltimus finis non est ppter finem: sed in quibusdā actiones sunt vltimus: vt pz per phm in pmo ethicorum: g non oia homo agit ppter finē. Pr. tunc vr homo agere ppter finē quādo deliberat: sz multa hō agit absq deliberatiōe de quibus etiam quandoq nihil cogitat: sicut cum aliquis mouet pedem vel manum alijs intentus vel fricat barbā: non g homo oia agit ppter finē. Sz ɔ. Omnia que sunt in aliquo genere deriuātur a pncipio illius generis: sed finis est pncipiū in operabilib⁹ ab hōie: vt pz per phm in scdo physicorum: g hōi ɔuenit oia agere ppter finē. Rñdeo dicendum q actionū que ab hoie aguntur ille sole ppie dicuntur humane que sunt ppie hominis inquātum est homo. Differt autē homo ab alijs irrationalib⁹ creaturis in hoc q ē suorum actuū dñs. Unde ille sole actiones vocantur ppie humane quarum homo est dñs. Est aūt homo dñs suorum actuum per rationē et voluntatē. vnde et liberū arbitriū esse dicit facultas voluntatis et rōnis. Ille g actiones ppe humane dicūtur que ex voluntate deliberata pcedunt: sique autē alie actiones hōi conueniant possunt dici quidē hois actiones: sz nō ppie humane cū nō sint hois inquantū ē hō. Manifestum est autē q omnes actiones que pcedunt ab aliqua potentia causantur ab ea fm rōnem sui obiecti. Obiectum autē voluntatis est finis et bonum. vnde oportet q omnes actiones humane ppter finem sint. Ad pmū g dicendū q finis et si sit postremus in executione: est tñ primus in intentione agētis: et hoc modo hz rationē cause. Ad secundū dō. q siqua actio humana sit vltimus finis oportet eam esse voluntariam: al's nō esset humana vt dictum est. Actio autē aliqua dupliciter dicitur voluntaria. Uno modo quia imperatur a voluntate: sicut ambulare vel loqui. Alio modo quia elicitur a voluntate: sicut ipm velle. Impossibile aūt est q ipse actus a voluntate elicitus sit vltimus finis. nā obiectū voluntatis ē finis sicut obiectū visus est color. Unde sicut impossibile est q primū visibile sit ipm videre quia oē videre est alicuius obiecti visibilis: ita est impossibile q pmum appetibile qd est finis sit ipsum velle. Unde relinquitur q siqua actio humana sit vltimus finis q ipsa sit imperata a voluntate: et ita ibi aliqua actio hominis ad minus ipm velle est ppter finem. quicquid g hō faciat verū est dicere q homo agat ppter finē etiā agendo actionē que est vltimus finis. Ad tertiū dō. q huiusmodi actiōes nō sūt ppie humane: qz non pcedunt ex deliberatione rationis que est pprium pncipium humanorum actuum: et ideo hñt quasi finē imaginatum non autē per rationem prestitutum.

Ad secundum sic proceditur. Uidetur q agere ppter finem sit pprium rōnalis nature. hō enī cui⁹ ē agē ppr finē nūq agit ppr finē ignotum sed multa sunt que non cognoscūt finem: vel qz oio carent cognitione sicut creature insensibiles: vel non apphēdunt rationē finis sicut bruta aialia. vr g pprium esse rationalis creature agere propter finem. Pr. agere ppter finem ē

a 2

Cat. 33. First question

THE PRIVATE AND THE PUBLIC GOOD

Johannes Versor (d. c. 1485) **and Antonius Andreae**

Quaestiones super libros ethicorum Aristotelis. [With commentaries on:] Oeconomica. Politica. [And with:] Antonius Andreae. Quaestiones super XII libros Metaphysicae

Cologne: Quentel, 1491–92; Leipzig: Wolfgang Stöckel, c. 1491
Annotated by Henricus Doring (Thuring) (15th–16th century)

This a collection of four fifteenth-century printed books annotated by a contemporary reader. The first three books are commentaries on Aristotle's *Nicomachean Ethics* and on his *Politics* by Johannes Versor, a French Dominican and philosopher and rector of the University of Paris who also commentated on Aristotle's natural philosophy (cat. 15). This volume, a collection made at the time of its printing, gathers together Versor's most important works: those devoted to man's public and private behavior. They have not been well appreciated by scholars, yet they are rich and valuable for the study of the transmission of Aristotle's ethics. At the dawn of the rise of states, Versor confronts the tension between the individual and the common good. This copy is heavily annotated by a single scholar, Henricus Doring (Thuring). His name is preserved in the notes, along with the price he paid in 1491 for this book: the considerable sum of 1 gulden and 10 new groschen. Densely annotated fifteenth-century printed books do not survive in great numbers; they are rarer still if they identify a specific reader in time and place.

Cat. 34. *Politics*, book I; the main text is in the larger font and Versor's commentary in the smaller font, with Henricus Doring's annotations in the margins

Liber primus yconomicoꝝ

Conomica ꝛ politica diffe rũt non soluz sicut domus et ciuitas. hec em̃ subiecta sunt eis. Ueruz etiam quia politica ex multis principi bus est. Jconomica vero monarchia. Artium q̃m alique sunt distincte. ꝛ nõ eiusdẽ est facere ꝛ vti eo q̃d factum est. puta vt lira ꝛ fistula. Politice vero est ciuitatẽ ab initio bene ↄstituere. ꝛ ea existẽte bene vti Patet etiã ꝙ yconomice sit domũ atq3 possessiones acquirere atq3 vti eis Ciuitas q̃dem igitur est domoꝝ pluralitas. est etiã prediorum ꝛ possessionum abũdans ad bene viuẽdum. Palam em̃ est quando nequierit hoc habere dissoluitur ciuitas ꝛ cõmunicatio Amplius autẽ q̃m huius causa ↄuenit. eius aũt causa vnũquodq3 est q̃d factuz est. ꝛ in substantia eius est hoc existens. Quare quidem palam ꝙ yconomica p̃or est politica nihilominus eius opus. Pars em̃ ciuitatis domus est: Uidenduz de yconomica quid sit opus eius

Eterminata parte prima moralis phie. nunc dicẽdũ est de secũda parte eiusdem. Et quia pars ista cum sequenti nũc plerosq3 latet ꝛ rara est. ideo vis principalior p̃mittetur ad q̃dam dubia. que iuxta capituloꝝ sententiã siue exigentiã interserũtur: ex quib9 relucissime poterit sub ↄpen

dio effectus pene totalis ↄprehendi ¶ Circa exordium igit' correspõdes huic parti secũde: (que est de bono ↄmuni domestico. ꝛ cuius sunt domestice. vt patuit in principio ethicoꝝ) est aduertendũ ꝙ iste liber totalis diuidit' in duos libros partiales. In p̃mo determinat de partib9 yconomice. ꝛ de cõitate viri ꝛ mulieris. que d' nuptialis. et de ↄmunitate parentũ ꝛ filioꝝ. que d' paterna. et dñoꝝ ad seruos. que d' dispotica. In secũdo v'o libro ↄplet illa que minus ↄplete st' in p̃mo dicta Primus liber subdiuidit' in duos tractatus. In scdo determinat de spẽb9 yconomice capiendo yconomicã p habitu quo quis domũ suã scit bene regere. Primus aũt tractat9 sex ↄtinet capl'a. In primo phus enumerat partes yconomice. s'm quas due erũt p̃tes v'les domus. scil3 hõ ꝛ possessio. per hoiem vir ꝛ mulier intelligũtur: sed p possessionẽ intelligitur possessio naturalis ꝛ artificialis Artificialis possessio dicit' sicut pecunie ꝛ vestes. Naturales vt agri prata Sed p̃tes materiales st' vir ꝛ vxor ꝛ bos arans. Et p̃ma cura yconomi esse debet de ↄiuge. et q̃tuz ad ea que st' extra domũ curã gerere d3 de agricultura. Prima aũt ꝛ necc̃aria possessio yconomi est ad seruũ. que ẽ possessio aiata. Et talis est duplex. sc3 curator. qui h3 p̃cipere. ꝛ opator q̃ h3 obedire. Huius aũt yconomice q̃tuor st' spẽs. sc3 acq̃sitiua. custoditiua. vsualis: ꝛ ordinatiua Deb3 em̃ habere prudentiã ad acq̃rendũ necessaria. custodiendũ expẽdẽda. ꝛ ordinãdũ domũ ꝛc et sic dom9 yconomici bñ stabit. Acq̃sitiua dispõ est vt res empte cito soluent'. ꝛ vendite res in custodia ad debitũ t'ps reseruent': Subiectũ aũt huius libri est hõ inq̃tum est p̃s dom9. et de illo melius patebit in dubio q̃rto. Liber aũt p̃ns precedit politicoꝝ. pro tanto quia ille tractat de ↄmunitate. que est ciuitas. p̃sens aũt liber de ↄmunitate domus. que est pars p̃oris cõmunitatis. sed pars est prior suo toto. et quia de partibus prior est speculatio. ideo rationabiliter presens liber precedit politicu3: Et differt scientia libri presentis a scientia libri politicorum cum quo etiam ↄuenit in hoc. Nam scientia presentis libri est de ↄstitutione dom9 que est pars ciuitatis. qua ↄstructa scie' quomõ debeat regi. Sic etia3 scientia politicoꝝ est de cõstructõe ciuitat'. q̃ ↄstruacta scie' q̃uo debeat regi Differũt tamen in hoc. quia politicoꝝ tractat de principibus ꝛ tota familia ciuitatis alicuius. Sed presens liber solum tractat de vno vnam domũ regentẽ sc3 ipso yconomo. Sic igitur patet intentio philosophi quo ad primum capitulum.

Ad euidentiã autez meliorem huius capituli erũt vndecim dubia. Quorũ primũ est hoc. Cur phs nõ absolute ostendebat q̃d sit yconomica: sed hoc facit ↄparando ea3 ad politicã cu3 tñ absoluta ↄsideratio precedere debet respectiuã Ad hoc dicitur ꝙ yconomica est pars politice. modo pars habet se respectiue ad suũ totum. Et hoc considerando partem formaliter. Materialiter vero p̃s est q̃dam res absoluta que q̃uis nõ sit sine illo respectu. potest tñ absq3 illo ↄsiderari. et sic habet in se aliquid per quod ab alio distinguitur. Et q' phs ostendit quid sit yconomica. ideo rõnabiliter interuenit ↄpatio ¶ Dubitat' secũdo quare nõ posuit mediam inter yconõicã ꝛ monasticã. sicut posuit int' yconõicã ꝛ politicã

Cat. 34. Beginning of the text and commentary of the *Economics*, now attributed to one of Aristotle's pupils or to Theophrastus, but for centuries transmitted as part of Aristotle's works

35

LEIPZIG, ARISTOTLE, AND THE REFORMATION

Sammelband of five Aristotelian works and commentaries

Annotated by Arnold Wöstefeld (1477–1540)

Aristotle and John Argyropoulos
(translator, 1415–1487)

Opus ... de moribus ad Nicomachum ... a Joannae Argyropilo ... traductum

Leipzig: Martin [Landsberg], 1511

Virgilius Wellendorfer
(late 15th century–16th century)

Moralogium (ex Aristotelis Ethicorum commentatorumque lecturis) ... authorisatum

Leipzig: Wolfgang Stöckel, 25 September 1509

Virgilius Wellendorfer
(late 15th century–16th century)

Oecologium ex duobus Aristotelis Oeconomicorum libellis accumulatum. Conclusiones centum et quattuor: Ac nove traductionis [Leonardo Arethini] ... complectens

Leipzig: Wolfgang Stöckel, 1511

Erasmus Friesner de Wonsiedel
(d. 1498)

Exercitium totius veteris artis. Predicabilium Porphirii: Predicamentorum et perihermenias Aristotelis; Sex principiorum Gilberti libros

Leipzig: Jacob Tanner, 1511

Erasmus Friesner de Wonsiedel
(d. 1498)

Exercitium physicorum Aristotelis octo libros ... complectens

Leipzig: Wolfgang Stöckel, 1511

Leipzig was an important economic and cultural center in sixteenth-century Germany, and home to one of the foremost universities of the day. This book gives us a very rare snapshot of an age. It is a Sammelband (a German term meaning several works gathered in a bound volume), consisting of five Aristotelian works and modern commentaries, in which every part speaks to the intellectual ferment running through Leipzig on the eve of the Protestant Reformation. The commentators were all University of Leipzig scholars, and the annotator, who was responsible for gathering the works together, was the well-known Leipzig humanist, philologist, and book collector Arnold Wöstefeld (1477–1540). The binding—blind-tooled pigskin over wooden boards—was certainly made contemporaneously with the creation of the volume.

Cat. 35. Beginning of the *Nicomachean Ethics*, translated by John Argyropolous

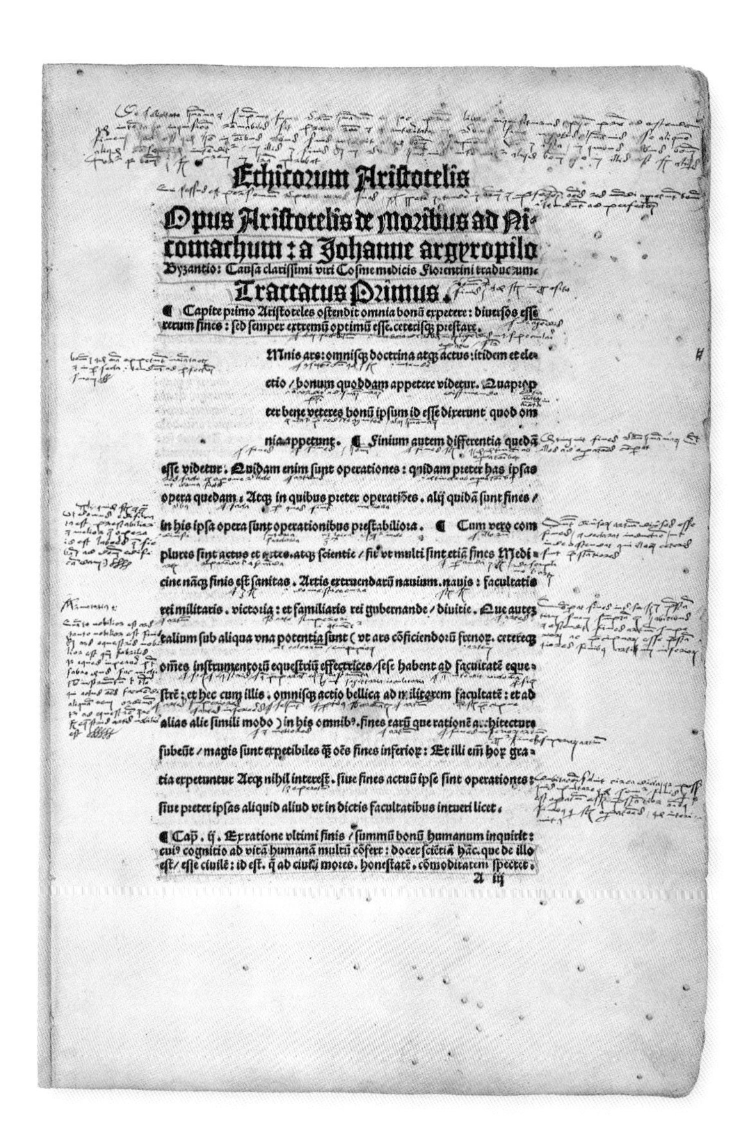

Ethicorum Aristotelis

Opus Aristotelis de moribus ad Nicomachum : a Johanne argyropilo Byzantio: Causa clarissimi viri Cosme medicis Florentini traductum.

Tractatus Primus.

¶ Capite primo Aristoteles ostendit omnia bonū expetere : diversos esse rerum fines : sed semper extremū optimū esse. ceterisq3 prestare.

Omnis ars: omnisq3 doctrina atq3 actus : itidem et electio / bonum quoddam appetere videtur. Quapropter bene veteres bonū ipsum id esse dixerunt quod omnia appetunt. ¶ Finium autem differentia quedā esse videtur. Quidam enim sunt operationes : quidam preter has ipsas opera quedam. Atq3 in quibus preter operatiões . alij quidā sunt fines / in his ipsa opera sunt operationibus prestabiliora. ¶ Cum vero complures sint actus et artes. atq3 scientie / fit vt multi sint etiā fines Medicine nāq3 finis est sanitas. Artis extruendarū navium. navis : facultatis rei militaris. victoria : et familiaris rei gubernande / divitie. Que autem talium sub aliqua vna potentia sunt (vt ars cōficiendorū frenor̄. ceterecq3 omēs instrumentorū equestriū effectrices / sese habent ad facultatē equestrē : et hec cum illis. omnisq3 actio bellica ad militarem facultatē : et ad alias alie simili modo) in his omnib9. fines earū que rationē a[illegible]chitecture subeūt / magis sunt expetibiles q̄ oēs fines inferior̄ : Et illi em̄ hor̄ gratia expetuntur Atq3 nihil interest. sive fines actuū ipse sint operationes : sive preter ipsas aliquid aliud vt in dictis facultatibus intueri licet.

¶ Cap̄. ij. Ex ratione vltimi finis / summū bonū humanum inquirit : cui9 cognitio ad vitā humanā multū cōfert : docet sciētiā hāc. que de illo est / esse civilē : id est. q̄ ad civi[illegible] mores. honestatē. cōmoditatem spectet.

A iij

Wöstefeld was a student of Aristotelian texts. Typically for his age and for people with sufficient means, he sought to buy books in order to have his own study copies. His was the largest private library in Leipzig at the time, containing about five hundred items. In this Sammelband, Wöstefeld included works and commentaries on ethics, economics, logic, and physics. The *Nicomachean Ethics* is bound at the beginning of the volume, in the celebrated Latin translation of the Byzantine Aristotelian John Argyropoulos. During the Renaissance, this text began to rival Grosseteste's version and related commentaries by Leipzig scholars. Only a few years after the creation of this volume, Leipzig University professors staged the most important debate of the early Reformation, the disputation of 1519, during which Martin Luther and Philipp Melanchthon took on defenders of the Catholic Church and debated the critical issues: papal indulgences, the legitimacy of the papacy, and the nature of free will and divine grace. The disputation marked Luther's final break with the Church.

36

A BYZANTINE COMMENTATOR ON THE ETHICS

Eustratius (c. 1050/60–c. 1120)

Commentaria in libros ... De moribus ad Nicomachum **[in Greek]**

Venice: Heirs of Aldus Manutius and Andrea of Asola, July 1536
Annotated by an unidentified Italian scholar (16th century)

Eustratius, twelfth-century metropolitan bishop of Nicaea, was one of the most important theologians at the Byzantine court of Alexius I Comnenus (r. 1081–1118). His commentary on the *Nicomachean Ethics* was the most influential of the Byzantine period, and the only one of its age to be published by the house of Aldus Manutius. So popular was Eustratius's interpretation of Aristotle that in the Renaissance he was called simply "The Commentator," a sobriquet normally applied to Averroes (and before him to Alexander of Aphrodisias). His exegeses of books I and VI provided an enormously useful service: rather than explaining particular words in the text, they offered a wide, all-encompassing interpretation, making Aristotle's ethics more relevant to Eustratius's age.

The conceptual nature of Eustratius's commentary must have been stimulating to the sixteenth-century Italian annotator who writes in Greek in the margins of many pages in this copy. We can see from his notes that he is proposing an alternative interpretation of Aristotle. While Eustratius's commentary had highlighted the normative nature of ethics—that is, how we construct sets of behavioral rules by which we can tell what is right from what is wrong—the anonymous owner of this copy is much keener to see ethics as "eudemonic," oriented towards identifying what makes us happy and the means to happiness. This book was once owned by Giacomo Soranzo (1686–1761), a Venetian senator from an illustrious dynasty of doges, humanists, and men of letters. It was then acquired by Joseph Smith (1682–1770), British consul in Venice from 1744 to 1760, a keen and discerning art collector and connoisseur.

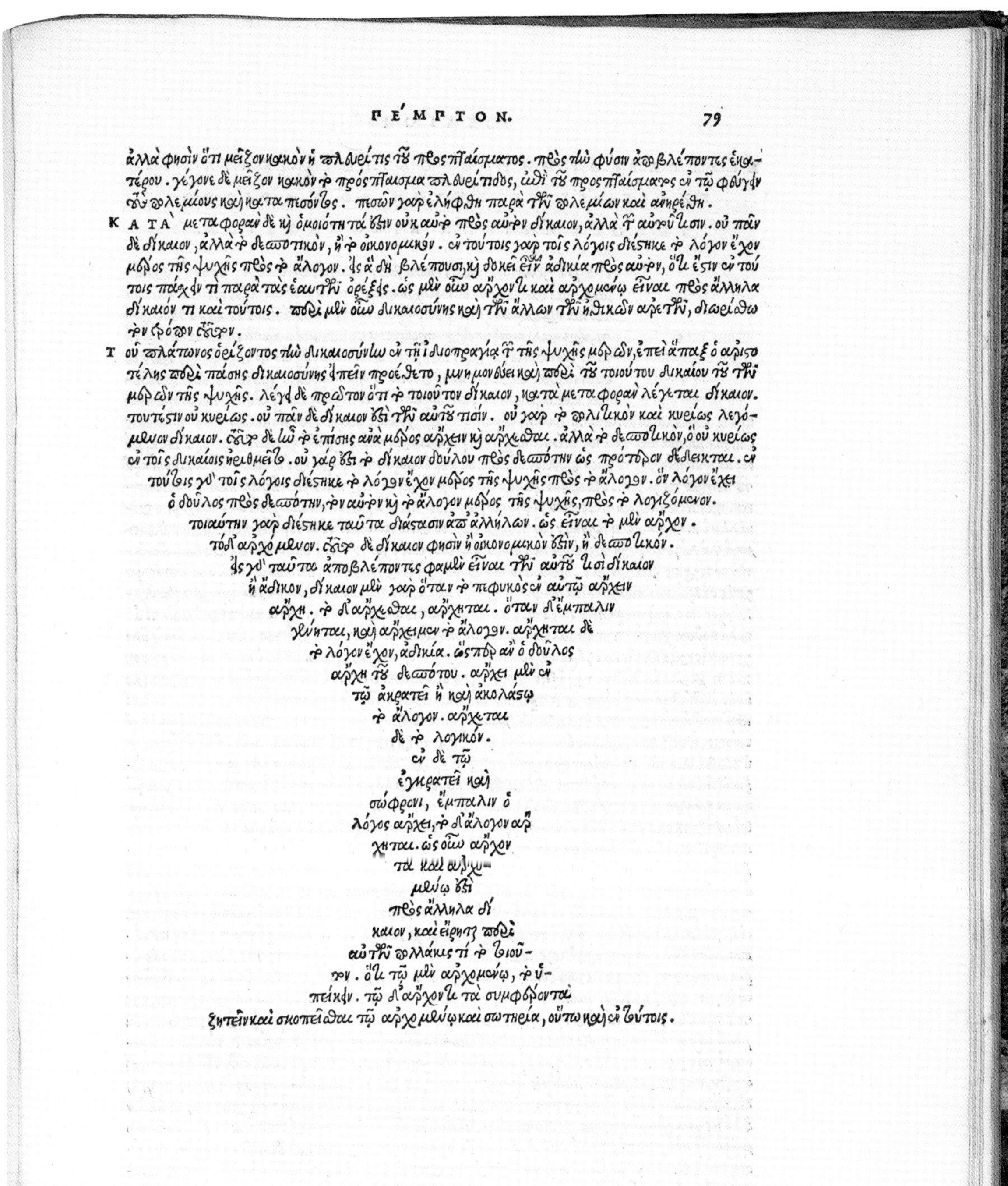

ἀλλά φησιν ὅτι μεῖζον κακὸν ἡ πλευρῖτις τοῦ προσπταίσματος. πρὸς τὴν φύσιν ἀποβλέποντες ἑκα-
τέρου. γέγονε δὲ μεῖζον κακὸν τὸ πρόσπταισμα πλευρίτιδος, ὡδὶ τοῦ προσπταίσματος ἐν τῷ φεύγειν
τοὺς πολεμίους καὶ καταπεσόντος. πεσὼν γὰρ ἐλήφθη παρὰ τῶν πολεμίων καὶ ἀνῃρέθη.

ΚΑΤΑ΄ μεταφορὰν δὲ καὶ ὁμοιότητά ἐστιν οὐκ αὐτὸ πρὸς αὑτὸν δίκαιον, ἀλλὰ τῶν αὐτοῦ τισιν. οὐ πᾶν
δὲ δίκαιον, ἀλλὰ τὸ δεσποτικόν, ἢ τὸ οἰκονομικόν. ἐν τούτοις γὰρ τοῖς λόγοις διέστηκε τὸ λόγον ἔχον
μέρος τῆς ψυχῆς πρὸς τὸ ἄλογον. εἰς ἃ δὴ βλέπουσι, καὶ δοκεῖ εἶναι ἀδικία πρὸς αὑτόν, ὅτι ἔστιν ἐν τούτοις πάσχειν τι παρὰ τὰς ἑαυτῶν ὀρέξεις. ὡς μὲν οὖν ἄρχοντι καὶ ἀρχομένῳ εἶναι πρὸς ἄλληλα
δίκαιόν τι καὶ τούτοις. περὶ μὲν οὖν δικαιοσύνης καὶ τῶν ἄλλων τῶν ἠθικῶν ἀρετῶν, διωρίσθω
τὸν τρόπον τοῦτον.

Τ οῦ πλάτωνος ὁρίζοντος τὴν δικαιοσύνην ἐν τῇ ἰδιοπραγίᾳ τῶν τῆς ψυχῆς μερῶν, ἐπεὶ ἅπαξ ὁ ἀριστο-
τέλης περὶ πάσης δικαιοσύνης εἰπεῖν προέθετο, μνημονεύει καὶ περὶ τοῦ τοιούτου δικαίου τοῦ τῶν
μερῶν τῆς ψυχῆς. λέγει δὲ πρῶτον ὅτι τὸ τοιοῦτον δίκαιον, κατὰ μεταφορὰν λέγεται δίκαιον.
τουτέστιν οὐ κυρίως. οὐ πᾶν δὲ δίκαιον ἐστι τῶν αὐτοῦ τισίν. οὐ γὰρ τὸ πολιτικὸν καὶ κυρίως λεγό-
μενον δίκαιον. τοῦτο δὲ ἦν τὸ ἐπίσης ἀνὰ μέρος ἄρχειν καὶ ἄρχεσθαι. ἀλλὰ τὸ δεσποτικόν, ὃ οὐ κυρίως
ἐν τοῖς δικαίοις ἠριθμεῖτο. οὐ γάρ ἐστι τὸ δίκαιον δούλου πρὸς δεσπότην ὡς πρότερον δέδεικται. ἐν
τούτοις γοῦν τοῖς λόγοις διέστηκε τὸ λόγον ἔχον μέρος τῆς ψυχῆς πρὸς τὸ ἄλογον. ὃν λόγον ἔχει
ὁ δοῦλος πρὸς δεσπότην, τὸν αὐτὸν καὶ τὸ ἄλογον μέρος τῆς ψυχῆς, πρὸς τὸ λογιζόμενον.
τοιαύτην γὰρ διέστηκε ταῦτα διάστασιν ἀπ᾽ ἀλλήλων. ὡς εἶναι τὸ μὲν ἄρχον.
τὸ δ᾽ ἀρχόμενον. τοῦτο δὲ δίκαιον φησὶν ἢ οἰκονομικὸν ἐστὶν, ἢ δεσποτικόν.
εἰς γοῦν ταῦτα ἀποβλέποντες φαμὲν εἶναι τῶν αὐτοῦ τισὶ δίκαιον
ἢ ἄδικον, δίκαιον μὲν γὰρ ὅταν τὸ πεφυκὸς ἐν αὐτῷ ἄρχειν
ἄρχῃ. τὸ δ᾽ ἄρχεσθαι, ἄρχηται. ὅταν δ᾽ ἔμπαλιν
γίνηται, καὶ ἄρχει μὲν τὸ ἄλογον. ἄρχηται δὲ
τὸ λόγον ἔχον, ἀδικία. ὥσπερ ἂν ὁ δοῦλος
ἄρχῃ τοῦ δεσπότου. ἄρχει μὲν ἐν
τῷ ἀκρατεῖ ἢ καὶ ἀκολάστῳ
τὸ ἄλογον. ἄρχεται
δὲ τὸ λογικόν.
ἐν δὲ τῷ
ἐγκρατεῖ καὶ
σώφρονι, ἔμπαλιν ὁ
λόγος ἄρχει, τὸ δ᾽ ἄλογον ἄρχηται. ὡς οὖν ἄρχον-
τι καὶ ἀρχο-
μένῳ ἐστὶ
πρὸς ἄλληλα δί-
καιον, καὶ εἴρηται περὶ
αὐτῶν πολλάκις τί τὸ τοιοῦτον. ὅτι τῷ μὲν ἀρχομένῳ, τὸ ὑ-
πείκειν. τῷ δ᾽ ἄρχοντι τὰ συμφέροντα
ζητεῖν καὶ σκοπεῖσθαι τῷ ἀρχομένῳ καὶ σωτηρία, οὕτω καὶ ἐν τούτοις.

Cat. 36. Text arrangement conveying additional meanings, typical of pattern poetry

37

POLITICS, ETHICS, ECONOMICS

Franciscus Valentianus (16th–17th century)

In libros 10 Aristotelis ad Nicomacum ... Praefatio ad octo libros Politicorum Aristotelis

[Possibly France, c. 1590–1610]
Manuscript on paper

What is the purpose of life? What is good; what is happiness; what are magnificence, justice, and friendship? What is money and how does it work? Such questions were keenly discussed in the Renaissance, prompted by notions Aristotle had set forth in the *Nicomachean Ethics* and the *Politics.* The identity of the author of this manuscript is revealed by a stamp on the binding giving us a name and a place of origin: Franciscus from Valence. We can date the manuscript using the watermark—a distinctive image or pattern impressed in paper by each paper maker, serving as a trademark. The watermark in this manuscript indicates a date in the late sixteenth or early seventeenth century.

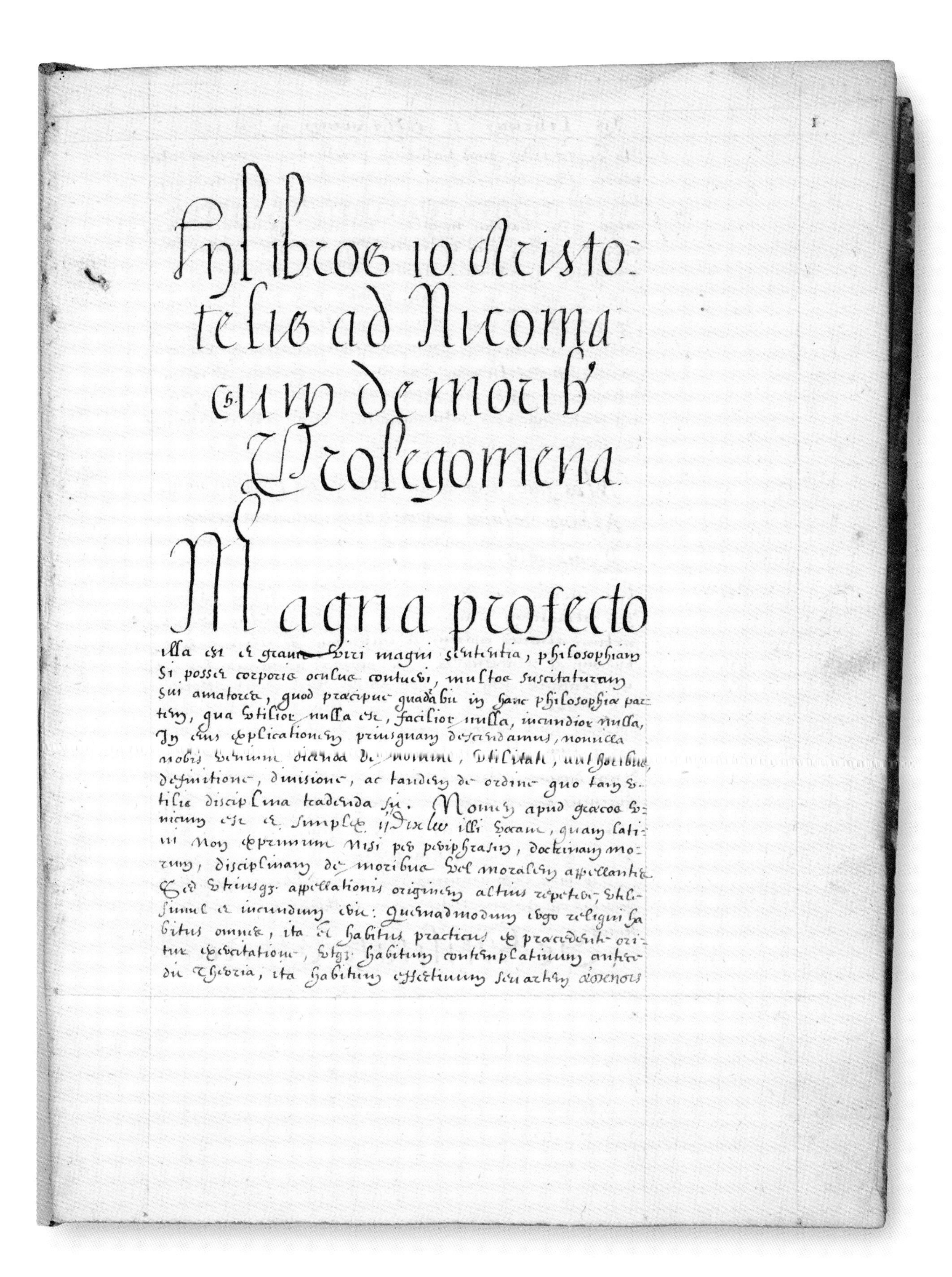

In libros 10 Aristotelis ad Nicomachum de moribus Prolegomena

Magna profecto illa est et gravis viri magni sententia, philosophiam si posset corporis oculis contueri, multos suscitaturam sui amatores, quod praecipue quadrabit in hanc philosophiae partem, qua utilior nulla est, facilior nulla, iucundior nulla. In cuius explicationem priusquam descendamus, nonnulla nobis breviter dicenda de nomine, utilitate, auctoribus, definitione, divisione, ac tandem de ordine quo tam utilis disciplina tradenda sit. Nomen apud graecos unicum est et simplex ἠθικὴ illi vocant, quam latini non exprimunt nisi per periphrasin, doctrinam morum, disciplinam de moribus vel moralem appellantes. Sed utriusque appellationis originem altius repetere utile simul et iucundum erit. Quemadmodum ergo reliqui habitus omnes, ita et habitus practicus ex praecedenti oritur exercitatione, utque habitum contemplativum appellamus Theoria, ita habitum effectivum seu activum [illegible]

Cat. 37. First page

finitum impossibile est moveri, nisi eandem magnitudinem semper
et non illud idem quod movetur, si contingeret. Quod quidem igitur si aliquid sit
eque velociter, necesse est finito finitum moveri manifestum est.
Accepta enim parte que mensurabit totum, in equalibus partibus tot quot
partes sunt per totum motum erit. Quare quoniam hee finite sunt et
quantitate unaquaque et tot modis omnes, et tempus utique erit finitum.
Tot enim modis erit hoc quantum tempus est partis multiplicate
secundum multitudinem partium. Sed et si non sit eque velociter, differt
nichil. Sit enim in quo est a b spatium finitum quod motum est
in infinito et tempus infinitum in quo sit c d. Si igitur necesse prius aliud alio motum esse. Hoc autem manifestum est quod in priori tempore et posteriori
aliud motum est. Semper enim in pluri aliud erit motum, sive eque
velociter mutet sive non eque velociter, et sive intendatur motus sive
remittatur sive maneat, nichil minus. Accipiatur igitur aliquod spatium quod
sit a e quod mensurabit a b. Hoc autem tam finitum in quodam est infinito
tempore. In infinito enim non potest esse. Omne enim in infinito est.
Et iterum aliud tantum si accipiamus quantum est a e necesse est in finito
tempore esse. Omne enim totum in infinito in sic accipiendi est.
Quoniam autem nulla est pars que mensuret totum. Impossibile enim
infinitum esse ex finitis et equalibus et inequalibus, eo quod men-
surantur finita multitudine et magnitudine ab
uno, sive equalia sive inequalia finita magnitudine.
Nichil minus autem spatium est finitum quantis quanta sunt a b men-
suratur, quare in finito tempore a b movebitur. Similiter
et in quiete. Quare neque si neque contingit unum ipsum possibile est semper idem

FURTHER READING

The Works of Aristotle

Jonathan Barnes, ed. *The Complete Works of Aristotle: The Revised Oxford Translation*. Princeton: Princeton University Press, 1984.

Edward Craig, ed. *Routledge Encyclopedia of Philosophy*. 10 vols. London and New York: Routledge, 1998.

Selected Individual Translations of Aristotle

J. L. Ackrill, trans. *Aristotle: "Categories" and "De Interpretatione."* Clarendon Aristotle Series. Oxford: Clarendon Press, 1963.

Ernest Barker, trans. *The Politics of Aristotle*. Oxford: Clarendon Press, 1948.

Jonathan Barnes and Anthony Kenny, eds. *Aristotle's Ethics: Writings from the Complete Works*. Rev. ed. Princeton: Princeton University Press, 2014.

Fred D. Miller Jr., trans. *Aristotle: "On the Soul" and Other Psychological Works*. Oxford World's Classics. Oxford: Oxford University Press, 2018.

C. D. C. Reeve, trans. *Aristotle: "Metaphysics."* Indianapolis: Hackett, 2016.

Robin Smith, trans. *Aristotle: "Prior Analytics."* Indianapolis: Hackett, 1989.

Robin Waterfield, trans. *Aristotle: "Physics."* Oxford World's Classics. Oxford: Oxford University Press, 1996.

Books about Aristotle

Jonathan Barnes. *Aristotle: A Very Short Introduction*. Oxford: Oxford University Press, 2001.

Jonathan Barnes, ed. *The Cambridge Companion to Aristotle*. Cambridge: Cambridge University Press, 1995.

Sarah Broadie. *Ethics with Aristotle*. New York: Oxford University Press, 1993.

Gail Fine. *On Ideas: Aristotle's Criticism of Plato's Theory of Forms*. New York: Oxford University Press, 1995.

David Furley, ed. *From Aristotle to Augustine*. Vol. 2 of *Routledge History of Philosophy*. London and New York: Routledge, 2006.

W. K. C. Guthrie. *A History of Greek Philosophy*. 6 vols. Cambridge: Cambridge University Press, 1981–2001.

Aristotle
De physica
France, 13th century
(detail of cat. 13, verso)

Thomas Heath. *A History of Greek Mathematics*. Oxford: Clarendon Press, 1960.

Thomas Heath. *Mathematics in Aristotle*. Oxford: Oxford University Press, 1949.

T. H. Irwin. *Aristotle's First Principles*. New York: Oxford University Press, 1990.

Jonathan Lear. *Aristotle: The Desire to Understand*. Cambridge: Cambridge University Press, 1988.

Armand Marie Leroi. *The Lagoon: How Aristotle Invented Science*. New York: Penguin, 2014.

Benjamin Morison. *On Location: Aristotle's Concept of Place*. New York: Oxford University Press, 2002.

Martha C. Nussbaum and Amelie Oksenberg Rorty, eds. *Essays on Aristotle's "De Anima."* New York: Oxford University Press, 1995.

Aristotle's Influence

Brian P. Copenhaver and Charles Schmitt, eds. *Renaissance Philosophy*. Oxford and New York: Oxford University Press, 1992.

A. C. Crombie. *Robert Grosseteste and the Origins of Experimental Science, 1100–1700*. Oxford: Clarendon Press, 1953.

Edward Grant, ed. *A Source Book in Medieval Science*. Cambridge, MA: Harvard University Press, 1974.

Dimitri Gutas. *Avicenna and the Aristotelian Tradition*. Leiden: Brill, 2014.

Dimitri Gutas. *Greek Thought, Arabic Culture*. London and New York: Routledge, 1998.

Moshe Halbertal. *Maimonides: Life and Thought*. Trans. Joel Linsider. Princeton: Princeton University Press, 2014.

Raphael Jospe. *Jewish Philosophy in the Middle Ages*. Boston: Academic Studies Press, 2009.

Robert Pasnau, ed. *The Cambridge History of Medieval Philosophy*. 2 vols. Rev. ed. Cambridge: Cambridge University Press, 2014.

Charles B. Schmidt and Quentin Skinner, eds. *The Cambridge History of Renaissance Philosophy*. Cambridge: Cambridge University Press, 1988.

Richard Sorabji, ed. *Aristotle Transformed: The Ancient Commentators and Their Influence*. London: Duckworth, 1990.

Richard Sorabji. *Philoponus and the Rejection of Aristotelian Science*. Ithaca: Cornell University Press, 1987.

Richard Sorabji. *The Philosophy of the Commentators, 20–600 AD: A Sourcebook*. Bristol: Bristol Classical Press, 2012.

PUBLICATIONS SPONSORED BY MARTIN J. GROSS AND THE MARTIN J. GROSS FAMILY FOUNDATION

David Hume, A Treatise on Human Nature, Modern Hebrew translation, edited by Mark Steiner. Jerusalem: Shalem Press, 2013.

Desire and Affect: Spinoza as Psychologist; Papers Presented at the Third Jerusalem Conference (Ethica 3), edited by Yirmiyahu Yovel. New York: Little Room Press, 1999.

Immanuel Kant, Critique of Pure Reason, translated into Modern Hebrew by Yirmiyahu Yovel. Israel: Hakibbutz Hameuchad – Sifriat Poalim, 2013.

Jewish Country Houses: The Lure of the Land, edited by Juliet Carey and Abigail Green. Forthcoming.

Jewish Treasures from Oxford Libraries, edited by Rebecca Abrams and César Merchán-Hamann. Oxford: Bodleian Library, University of Oxford, 2020.

Kant's Philosophical Revolution: A Short Guide to the Critique of Pure Reason, by Yirmiyahu Yovel. Princeton: Princeton University Press, 2018.

The Kennicott Bible. Oxford: Bodleian Library, University of Oxford, forthcoming.

Out of Exodus: The Evolution and Legacy of the Exodus Story, edited by Rebecca Abrams. Forthcoming.

Spinoza on Reason and the "Free Man": Papers Presented at the Fourth Jerusalem Conference (Ethica 4), edited by Yirmiyahu Yovel and Gideon Segal. New York: Little Room Press, 2004.

The Story of the National Library of Israel. Forthcoming.

Voltaire: The Martin J. Gross Collection in the New York Public Library, by Paul LeClerc, Martin J. Gross, and Stephen Weissman. New York: New York Public Library, 2008.